Amphibians & Reptiles
of
Sanibel & Captiva Islands, Flo[rida]

To the Ryan's,
Thanks for your interest in reptiles + amphibians. I hope you enjoy the book!

Charles LeBuff & Chris Lechowic[z]

Amphibians & Reptiles
of
Sanibel & Captiva Islands, Florida

*To the Ryans,
Thanks for your interest in reptiles + amphibians. I hope you enjoy the book!*

Charles LeBuff & Chris Lechowicz

Above: A subadult loggerhead sea turtle (*Caretta caretta*) grooms its shell by scraping it against a limestone rock outcrop at Gray's Reef National Marine Sanctuary. This reef formation is located in 19 meters (10.4 fathoms) of water offshore of Sapelo Island, Georgia, in the Atlantic Ocean. Loggerheads regularly visit such convenient sites to remove encrusting organisms that foul their carapaces. The very formation of this cleaning station is believed to have resulted because of long-term use by grooming marine turtles. *Michael G. Frick*

The cover: The extremely large female American crocodile (*Crocodylus acutus*) on the cover established a permanent territory and became a full-time resident of Sanibel Island, in 1979. This photograph captures her in a typical crocodile pose. The wide-open mouth while basking allows for thermoregulation. Even after capture and relocation to a state park 80 kilometers (50 miles) south of Sanibel Island she quickly returned. This crocodile remained a popular attraction on J. N. "Ding" Darling National Wildlife Refuge for more than 30 years until her death in 2010. *U.S. Fish and Wildlife Service*

On the back cover: A pair of pig frogs (*Lithobates grylio*). The male is to the left and is identifiable because of the large tympanum or ear disc posterior to its eye. This South Florida counterpart of the American bullfrog did not occur on Sanibel Island until 1952 when it was introduced to the Bailey Tract by wildlife refuge officials to help diversify wildlife prey. *Bill Love*

Amphibians & Reptiles
of
Sanibel & Captiva Islands, Florida

by

Charles LeBuff and Chris Lechowicz

*Featuring the photography
of Bill Love & Daniel Parker*

Foreword by C. Kenneth Dodd., Jr., Ph.D.

©2013, Charles LeBuff and Chris Lechowicz.

All rights reserved. No part of this book may be reproduced or transmitted in any form or by any means—graphic, electronic, or mechanical, including photocopying, recording, or any information storage or retrieval system—without written permission from the authors.

Designed by Charles LeBuff
Printed in China by Four Colour Print Group, Louisville, Kentucky.

Published by:
　Amber Publishing
　14040 - 101 Eagle Ridge Lakes Drive
　Fort Myers, Florida 33912-0711
　U.S.A.
　www.sanybel.com

In partnership with:

　Ralph Curtis Publishing
　Post Office Box 349
　Sanibel, Florida 33957-9998
　U.S.A.
　www.ralphcurtisbooks.com

First Printing, January 2014

Library of Congress Control Number: 2012923797

ISBN 978-0-9625013-4-0

For the inspirational pioneers of Florida herpetology, many of whom I knew and with whom I worked.

—Charles LeBuff

Dedicated to the people that decided to keep Sanibel as natural as possible and to generations entrusted to maintain it in the future.

—Chris Lechowicz

~CONTENTS~

Dedication	viii
Foreword	xii
Preface	xv
Acknowledgments	xviii
Introduction	xxii
List of Abbreviations	xxiii

PART I

Scope of the Book	1
Location	1
Geologic Formation	2
Climate	5
Land Use History	6
Hydrology and Water Management	8
Mosquito Control	11
Island Ecosystems	12
Species Recruitment	23

Environmental and Other Threats to the Island Herpetofauna	24
Early Herpetological Evidence	29
Collection Techniques	30

PART II

Annotated List of Species	35
Order Anura—The Frogs	41
Order Testudinata—The Turtles	75
Order Crocodilia—The Crocodilians	135
Order Squamata (Lacertilia)—The Lizards	149
Order Squamata (Serpentes)—The Snakes	183
Supplemental Species List	228
Glossary	239
Notes	251
Suggested Reading	257
Literature Cited	261
About the Authors and Photographers	265
Index, Scientific Names	269
Index, General	271

~FOREWORD~

On July 14, 1935, the eminent Florida naturalist Archie Carr wrote, "Today I believe it would be possible to cross Central Florida by car and at no time be in a situation where the calls of [*Anaxyrus*] *quercicus* [the oak toad] were not audible" (Carr 1940). Those days belong to the distant past, and the temporary pond-dwelling oak toad has disappeared from much of its formerly contiguous habitat throughout the state. A drive across Central Florida today would be met only by strip malls, traffic, and degraded remnants of Florida's magnificent natural environment. Although always an optimist, Archie no doubt would be saddened at the spectacle and the lack of peeps from oak toads.

From the 1920s to 1940s, naturalists were increasingly appalled at the destruction of natural Florida. The titles of their books tell of their sorrow: John K. Small's *From Eden to Sahara, Florida's Tragedy* (1929); W. S. Blatchley's *In Days Agone* (1932); Thomas Barbour's *That Vanishing Eden* (1944). In his book *Florida Wild Life* (1932), Charles Torrey Simpson entitled the last chapter "In Memoriam" in a lament of the devastation of Florida's biotic communities, noting vast changes from the time he began his studies 50 years previously. Most of these writers were botanists or wide-ranging naturalists, and they tended to focus on the plant communities, physical landscapes, and remnants of Native American presence, particularly shell mounds. Some authors referred anecdotally to the herpetofauna they observed, particularly alligators, rattlesnakes, and turtles, but none provided a comprehensive picture of the diversity of Florida's pre-European settlement amphibian and reptile fauna.

From 1773 to 1777, William Bartram, one of America's first great naturalists, trekked throughout the Southeast and made numerous observations on the amphibians and reptiles of Florida. He recorded accurate observations of many aspects of the behavior of reptiles, particularly snakes and alligators. Some of his observations were in error or seemed fantastical, but a number were later verified by biologists (Adler 2004). He noted the great abundance of alligators, but unfortunately we will never have a quantitative insight into the abundance of Florida's amphibians and reptiles prior to the wholesale loss of habitat that occurred in the 19th and 20th centuries. William Bartram never made it to Sanibel-Captiva, but then most naturalists tended to bypass these islands on Florida's southwest coast. For example, the coleopterist Willis Blatchley spent only a single day collecting shells and insects on Sanibel Island in 1913, and "found few insects except mosquitoes, which are here by millions."

There are few truly long-term studies of the biota of any single location in Florida. In 1958, however, Charles LeBuff began his lifelong observations of the flora and fauna of Sanibel and Captiva islands. Keeping meticulous records and notes of sightings and behavior of amphibians and reptiles for more than 50 years, Charles gained a unique understanding of an entire fauna as few biologists have ever been able to attain.

Charles' interest in Sanibel-Captiva extends far beyond the lives of the animals themselves. He clearly understands the maxim I try to instill in graduate students—that is, there are three facets to understanding why species presently occur where they do. First, one must know about their life history and evolution. Second, one must understand their physical and biotic environment. Third, one must become a historian and delve into the influences of land use and human history upon them. These prerequisites of exceptional natural history are clearly demonstrated in *Amphibians and Reptiles of Sanibel and Captiva Islands, Florida*.

In this book, Charles and co-author Chris Lechowicz discuss not only the life history of the amphibians and reptiles that occur today on Sanibel-Captiva, but also review records demonstrating historical loss and a decrease in diversity. The data are not based on hearsay, but on actual records, sightings, and preserved specimens methodically compiled over five decades. The authors are able to correlate changes in herpetofaunal diversity with changes in the biotic and physical landscapes of the islands resulting from both natural and human-induced causes. Change is natural in ecosystems, and coastal islands are often subject to dramatic and sudden storm-induced changes. The changes wrought by human activity, however, have been impressively dynamic on Sanibel especially during the latter part of the 20th century. In a previous book (2011), LeBuff assembled a photographic history of Sanibel-Captiva based on his work at "Ding" Darling National Wildlife Refuge. Coupled with this review of the islands' herpetofauna, I know of few such compilations where one can read about changes and simultaneously see them as they took place. *Amphibians and Reptiles of Sanibel and Captiva Islands, Florida* is thus more than a rendition of the herpetofauna of two Southwest Florida islands. It is about a sense of place.

Fully one-third of the world's amphibians are imperiled or in immediate danger of extinction. Scientists fear that nearly as many reptiles are similarly declining as human populations explode during an era of extinction termed the Anthropocene. To stem this tide, conservation, and indeed biological science, must begin with precise natural history information to mitigate past transgressions. Although natural history is often frowned upon by academic scientists and funding agencies, no short-term study or experiment could have set out to explore the intricate changes to Sanibel-Captiva that occurred over 50 years to the entire fauna. It is even more remarkable that a single natural historian (a person of nature and history) has been able to document such long-term changes.

Naturalists have a broad familiarity with different organisms and ecosystems and are inspired by a desire for knowledge and understanding of living creatures. Their knowledge invariably leads them to teach about and protect the source of that knowledge and inspiration. In the true sense of the term, Charles and Chris are naturalists of the 21st century, grounded in science but with keen eyes for the breadth of fauna and an appreciation of the inter-connectedness of living species. *Amphibians and Reptiles of Sanibel and Captiva Islands, Florida* is undeniably a foremost contribution to our knowledge of Florida's changing ecosystems and of the amphibians and reptiles that comprise such important components of its wildlife.

—C. Kenneth Dodd, Jr.
Department of Wildlife Ecology and Conservation, University of Florida

~PREFACE~

An egg-laden sea turtle, awash in the surf at the water's edge of primeval Sanibel Island, hesitates momentarily before making the decision to leave the water completely and ascend the low-sloped beach. This act is a prelude to a series of significant biological events that will follow, for she is the first reptile to use the habitat of this young island. At this point in time Sanibel Island is nothing more than a slightly emergent dry sandbar with its highest elevation scarcely more than a meter above the mean high-water line, having been recently created during the violent passage of a hurricane of cataclysmic proportions. Other than a few terrestrially adapted marine invertebrates such as ghost crabs (*Ocypode quadrata*), there are not yet any vertebrates dwelling on this accreting beach, although foraging or loafing ground-nesting seabirds have started to share the dry strip of sand and shell.

After resting briefly, this turtle laboriously hauls her heavy bulk out of the Gulf of Mexico and across the primitive beach. Soon, her mission is completed; her clutch of eggs has been covered, and she returns to the warm waters of the gulf. This lone turtle has irrevocably locked her species to the destiny of Sanibel Island.

Millennia have passed since this event, and this unique island has continued to change dynamically through time. The descendants of that first nesting sea turtle continue to visit Sanibel Island today.

Many factors have contributed to the formation of this island and the structure of its biological diversity. Even once-abundant native species of amphibians and reptiles, however, have been recently extirpated on Sanibel and Captiva islands. Exotic species, some totally undesirable, others relatively unnoticeable, have become moderately, even well, established in the island's ecosystems. Some are predatory and have preyed on and impacted the established native Florida herpetofauna stocks. This has caused a decline that may ultimately lead to the extirpation of other once-abundant species on these barrier islands.

The long-term changes in the herpetofaunal diversity of Sanibel and Captiva islands are important. The historical aspects of the initial indigenous amphibian and reptile species composition and population abundance are addressed herein, as are the avenues by which several mainland species may have reached the barrier islands. The stability in the abundance of native species that are well established, and the later introduction of both nonindigenous native amphibians and reptiles and the arrival of forms that are alien to the U.S., are also discussed.

The drastic changes to Sanibel Island's surface water aquifer are reviewed, and also the impacts of vegetational transition because of the invasion of exotic pest plants. Subsequent plant community alterations because of the efforts that have been undertaken to remove invasive non-native plants, along with real estate development of the barrier islands, have impacted the resident herpetofauna and enhanced the success of the exotic newcomers. All species of amphibians (13), turtles (17), crocodilians (2), lizards (15), and snakes (17) that have been documented on Sanibel and Captiva islands during the past 50 years are included in this overview and in their respective historical perspectives. A supplemental list includes additional species that have been collected at least once or reported but never validated on the island. Valid documentation can occur only with clear-cut photographic evidence or physical collection of a living specimen or a carcass. The status of each species over time is reviewed in an up-to-date, well-illustrated, and comprehensive annotated list of species.

Populations of species that were once part of a viable herpetofaunal community endemic to the islands are today stressed and are labeled Threatened or Endangered. The eastern indigo snake is considered to have disappeared from Sanibel and Captiva islands, although a lone individual was discovered on Captiva Island in 2012. We assume this snake originated on North Captiva. Sanibel's American alligators are undergoing survival pressures that were unimaginable just two decades ago. In the unusual case of the latter, the citizens of the City of Sanibel must bear the responsibility for the alligator's decline within the city limits, as over the short-term their local government demands the harvest and removal of immature alligators that exceed the arbitrary threshold size of 1.2 meters (m) (4 feet [ft]) in length. With such a policy in place it is inevitable that a viable breeding population of alligators will collapse on the island, and this may occur relatively soon. Taxpayers assume the costly burden via an ad valorem tax as property owners may be required to spend a large amount of money from the public treasury for what amounts to bounties. Currently, lizard-killing contractors are well paid to combat specific invasive reptiles, namely green iguanas and Nile monitors. Should the latter ever gain a foothold on the barrier islands, this voracious species may replace the alligator as Sanibel Island's ultimate predator other than man and his motor vehicles.

Little has been published on the herpetology of Sanibel and Captiva islands in more than two decades, although the amphibians and reptiles of these barrier islands have been closely monitored for more than a half century. The authors have combined their years of experience, fieldwork, and research on the islands and incorporated the anecdotal contributions of island naturalists to produce this contribution to the herpetology of Florida.

~ACKNOWLEDGMENTS~

We are vitally interested in the herpetology of Sanibel and Captiva islands. This group of animals is essential for a healthy diversity of indigenous wildlife species that help to set these unique islands apart from most other barrier islands. Although we worked independently and are separated in age by a generation, we chose to collaborate to produce this comprehensive work on the herpetofauna of these islands.

Charles LeBuff: I thank my wife, Jean, my partner for more than 56 years, for her continued encouragement when it was time to do my part in writing this contribution to the herpetology of Florida. We met in high school, and Jean was willing to share me with my herpetological interests from the beginning. This required much patience on her part. She often found escaped snakes in her cabinets; shared her home with turtles, a crocodile in the bathtub, and a 250-pound tortoise on the porch; and tolerated the strange-acting and stranger-looking herpetologists that often visited our home. Then there were the Sanibel summers, where for 32 years I was in the field *every* night, out on the beach nearly all night long, working with sea turtles.

Friends and other colleagues who contributed to this effort are too numerous to mention, but I will try.

In any aspect of herpetological fieldwork one usually has a circle of cohorts who have similar interests. I thank my brothers, Laban and Laurence, my sister, Natalie Kirkland, friends Richard Beatty, Warren Boutchia, Ralph Curtis, "Sonny" Marine, Ed Phillips, and George Weymouth for their help in the field in the very beginning. I also acknowledge the valuable contributions made by the scores of dedicated volunteers of Caretta Research, Inc. who aided me throughout the latter part of my herpetological career.

I especially thank retired wildlife biologist Richard L. Thompson, who once served as the project leader supervising the South Florida National Wildlife Refuges Complex including J. N. "Ding" Darling National Wildlife Refuge (JNDDNWR). A long-time friend, Dick kindly reviewed the first draft of the manuscript and made many valuable suggestions.

Although our paths crossed only once for two days in 1959, systematic herpetologist Douglas A. Rossman, Ph.D., retired professor of zoology at Louisiana State University in

Baton Rouge, Louisiana, agreed to read the final draft of this book. Doug provided wise council that helped us finalize the long-overdue manuscript, and we appreciate his input.

Likewise, C. Kenneth Dodd, Jr., Ph.D., now with the Department of Wildlife Ecology and Conservation at the University of Florida, reviewed our final manuscript. Ken first became acquainted with Sanibel and Captiva islands in 1984 when his official duties as a USFWS research zoologist brought him to the barrier islands to visit JNDDNWR. Through the years since, Ken's advice has been invaluable and he has contributed his expertise and editorial skills to help make this a much better book.

The current manager of JNDDNWR, Paul Tritaik, made the refuge's herpetological records available for review, for which we are thankful. Some of these were a little musty, having been housed in tightly packed refuge files for more than 50 years!

Steve Phillips, a former director of the SCCF, reviewed and shared with us his herpetological studies that were conducted during his tenure on Sanibel Island.

Interpretive naturalist, osprey specialist, former Sanibel mayor, and past nuisance alligator handler Mark "Bird" Westall helped us revisit and track the Sanibel Island Alligator Management program until it was suddenly abandoned—killed by a combination of tragedy and a silly leap toward political correctness.

Amanda Bryant, sea turtle biologist with the SCCF, graciously provided recent nesting data for loggerhead, green, leatherback, and Kemp's ridley sea turtles on Sanibel and Captiva islands.

Native-born Sanibellian Ralph Woodring provided an update on wild stocks of juvenile sea turtles he often encounters during his bait-shrimping operations around Sanibel Island, particularly Tarpon Bay.

Stanley Fox, Ph.D., Department of Zoology, Oklahoma State University (OSU), graciously searched his institution's collection archives and provided us with the catalog numbers of the series of Sanibel Island amphibians and reptiles that LeBuff furnished more than 50 years ago to the late Jack McCoy, then a graduate student at the university.

Susie Holly, book and magazine editor, as well as owner of MacIntosh Books and Paper on Sanibel, copy edited the final manuscript despite her profound phobia of snakes.

Betty Anholt, author, eminent Sanibel Island historian, and naturalist, proofread our final draft. We appreciate her careful attention to this project.

Chris Lechowicz: I would like to thank Nicole, my wife of 17 years, for all of her patience and support of my "herp"-related endeavors that often resulted in untimely absences from the household. I also thank my two sons, Merrick and Graham, for inspiring me to work harder at completing my goals.

Thank you to Brian Weber and Ron Humbert for taking my lifelong interest in herpetology to levels unimaginable in my youth. I owe a debt of gratitude to the Chicago Herpetological Society (CHS) for mentoring my early adulthood with much knowledge, professional guidance, and countless friendships that will last a lifetime.

I would also like to thank Erick Lindblad, executive director of SCCF, for his ongoing support and confidence in the research being conducted by the Wildlife Habitat Management Program at SCCF. Erick understands the importance of amphibians and reptiles in the ecosystem and the work we do with them.

Special thanks goes out to all the interns, interagency biologists (JNDDNWR and the City of Sanibel staff), and friends who helped with all the recent fieldwork needed to survey the islands—specifically, Amanda Bryant, Cara Faillace, Vinny Farallo, Sam Fuller, Shane Johnson, Joel Caouette, Heather Porter, Toby Clark, Victor Young, and Bill Love. A very special thanks goes to the Orianne Society for partnering with SCCF and getting the Pine Island Sound Eastern Indigo Snake Project off the ground. Last but not least, a heartfelt thank you to Bill Love and Daniel Parker for providing the majority of the photographs used in this book.

—Charles LeBuff
—Chris Lechowicz

~INTRODUCTION~

In 1958 the senior author was selected to fill the second permanent position at the Sanibel National Wildlife Refuge (SNWR). He was to serve as the refuge's wildlife biologist for the next 32 years. When he entered on duty in January 1959, because of his interest and background in Florida herpetology, the refuge manager instructed that whenever the opportunity arose he was to collect and catalog those species of amphibians and reptiles that occurred within the refuge proclamation boundary. Over time, a list of the Sanibel Island herpetofauna was compiled for the refuge (LeBuff 1959). This fieldwork and related conservation efforts spanned a period of 48 years.

In 1967, SNWR was renamed to honor the lifetime achievements of the late conservationist/political cartoonist Jay Norwood "Ding" Darling, who had wintered on Captiva Island between 1934 and 1960. Mr. Darling (1876-1962) helped spearhead the establishment of the refuge that now bears his name.

In 1959 LeBuff entered into a cooperative arrangement with the late Clarence J. McCoy, Jr., who at the time was a graduate student at Oklahoma State University (OSU) in Stillwater. A series of amphibians and reptiles from Southwest Florida, including a few species from Sanibel Island, were collected and shipped to McCoy. These were cataloged in the OSU collection of vertebrates. Where appropriate, we cite the OSU catalog numbers for that small collection of species from Sanibel Island that are represented in the annotated list of species that follows. A voucher collection of all amphibians and reptiles that were taken on Sanibel Island between 1959 and 1969 was once housed in the biology department of Cypress Lake High School, Fort Myers, Florida, but these have since been discarded.

Early field compilations, or published lists compiled by others, relative to the herpetofaunal forms that are known to range into, or that may otherwise occur exclusively in southern Florida as subspecies, have not always included verification of a species' actual presence on the chain of barrier islands. Carr (1940), Duellman and Schwartz (1959), and Truitt (1962), are early contributions that do not, or only incompletely, discuss the insular populations of these animals. Carr listed the counties where the various species had been collected, and a paucity of records within the range of each species at that time is evident. The mainland geographical range of any particular species does not justify including that species as indigenous to Sanibel and Captiva islands.

~List of Abbreviations~

BT	Bailey Tract
CCCL	Coastal Construction Control Line
CEPD	Captiva Erosion Prevention District
CHES	Charlotte Harbor Estuarine System
CHS	Chicago Herpetological Society
COS	City of Sanibel
CROW	Clinic for the Rehabilitation of Wildlife
DOR	Dead on road
FGFWFC	Florida Game and Freshwater Fish Commission
ESA	Endangered Species Act
FFWCC	Florida Fish and Wildlife Conservation Commission
FLMNH	Florida Museum of Natural History, Gainesville
GPS	Global Positioning System
IICA	Inter-Island Conservation Association
JNDDNWR	J. N. "Ding" Darling National Wildlife Refuge
LCMCD	Lee County Mosquito Control District
LNWR	Loxahatchee National Wildlife Refuge
MSL	Mean sea level
OSU	Oklahoma State University
PIT	Passive Integrated Transponder (tag)
RECON	River, Estuary and Coastal Observing Network
SCCF	Sanibel-Captiva Conservation Foundation
SILS	Sanibel Island Light Station
SNWR	Sanibel National Wildlife Refuge
SSAR	Society for the Study of Amphibians and Reptiles
STSSN	Sea Turtle Stranding and Salvage Network
SWFRAA	Southwest Florida Regional Alligator Association
UF	University of Florida
USFWS	U.S. Fish and Wildlife Service
USGS	U.S. Geological Survey

~PART I~

THE SCOPE OF THIS BOOK

Through all stages of this book's development, it has been our goal to make this work a meaningful and special contribution to the natural history of Sanibel and Captiva islands, and to the herpetology of Florida. We have strived to create a readable yet semi-technical treatise, one that contains biologically accurate text, the content of which directly benefits the scientific community. Furthermore, we have included a mix of anecdotal and historical information applicable to this area's herpetology, and we have provided photographs of the highest quality to complement the species accounts. We want our readers to discover a completely new set of facts about Sanibel and Captiva islands, and for the information we provide to allow for a level of understanding that exceeds a casual or cursory exploration of the herpetology relative to these islands. Our intent is to be comprehensive, rather than produce a purely popular summary and guide. We expect this book to remain an important piece of island-related literature for many years to come.

    Herpetology, the study of amphibians and reptiles, is one of the only zoological sciences where a student can get a personal hands-on involvement in the field with these unique animals. Although it is not entirely dissimilar to the other biological sciences, herpetology is unique in that it offers windows of opportunity for the scientist and layman alike to join forces and become intimately one-on-one with these organisms. The long-term sea turtle conservation effort on Sanibel and Captiva islands is an excellent case in point. In activities of this nature, it is the qualified laymen, naturalists, and "citizen scientists" who are the key to the operational success of such programs.

LOCATION

Sanibel and Captiva islands are located in Lee County on the Southwest Florida coast 170 kilometers (km) (106 miles [mi]) south of Tampa. Sanibel is an atypical barrier island in that it is geographically oriented east-west rather than north-south, and is not situated parallel to the mainland as is typical for barrier islands along the Florida peninsula. Three bodies of water—the gulf, Pine Island Sound, and San Carlos Bay—surround Sanibel Island.

At its closest point, the island is positioned 2.6 km (1.6 mi) offshore in the Gulf of Mexico from the mouth of the Caloosahatchee River[1]. The waters connected to this river form part of a complex regional estuary that includes the coastal waters from extreme southern Sarasota County, all of Charlotte County, all of Lee County, and extreme northern Collier County (see Map 5, below). This area is collectively known as the Charlotte Harbor Estuarine System (CHES) (McPherson et al, 2007). Historically, the Caloosahatchee drained the interior of the Florida peninsula. The river was channelized in the late 19th century and today connects directly with Lake Okeechobee through a series of navigable locks.

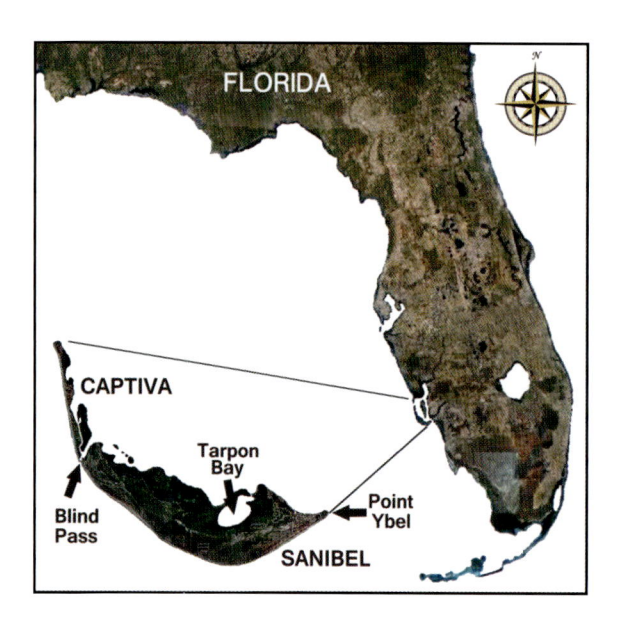

Map 1. Sanibel and Captiva islands are situated on the Gulf of Mexico seacoast in Southwest Florida, a region rich in herpetological diversity.

Captiva Island, from which Sanibel Island's geological origins evolved, is a more ordinary barrier island in that it is geographically positioned in a north-south orientation. This narrow island is situated between Pine Island Sound and the Gulf of Mexico.

Sanibel Island has been recognized since the 16th century as a unique natural feature of Southwest Florida. George R. Cooley (1955) summed this up when he described Sanibel Island thusly, "... *the island has the appearance of many sections of South Florida. Yet it is not exactly the same as any region of the tropics or of Florida, historically, geographically, geologically, or botanically.*"

Sanibel Island's reputation and uniqueness were further recognized in 2011 when international travel guru Arthur Frommer wrote that Sanibel Island was his number-one choice for a travel destination *in the world*. Among the reasons he cited was the existence of JNDDNWR.

GEOLOGIC FORMATION

Geologically, Sanibel Island is about 6,000 years old, some thousands of years younger than Captiva Island. Sanibel Island was formed from biogenic substrate, which was transported via littoral currents, or scoured from the bottom of the Gulf of Mexico and heaped longitudinally as a result of unrecorded post-Pleistocene cataclysmic 500-year storm events. The island formed as an ever-enlarging, slowly accreting appendage, a southeastward projection arching from Captiva Island. Over time, Sanibel Island became a large crescent-shaped barrier island with a maximum natural elevation of 4 meters (m) (13 ft) above mean sea level (see Map 1). Changes in sea level also helped shape the modern island (Missimer 1973). At present, Sanibel Island has 19.3 km (12 mi) of Gulf of Mexico frontal beach, nearly 3.2 km (2 mi) of bay beach along the shore of San Carlos Bay, and many kilometers of an outer discontinuous shoreline of frequently tidally overflowed red mangrove-dominated keys that meander into the waters of Pine Island Sound.

Historically, Captiva Island was delineated at its southern end by Blind Pass, which periodically separates Captiva from Sanibel, and Captiva Pass, a northern inlet that once separated Captiva from LaCosta[2] Island. In 1921 tidal surge from a major hurricane[3] overwashed the narrow island and broke Captiva into two islands by creating Redfish Pass. This major inlet now separates Captiva from Upper (or North) Captiva. Modern Captiva is 8 km (4.9 mi) in length and slightly more than 0.5 km (0.3 mi) across at its widest point. Until recently, Sanibel and Captiva were joined together after the Blind Pass opening had migrated and closed at the Blind Pass Bridge. This inlet's position fluctuates and is discussed in detail under Island Ecosystems.

Aerial photograph of western Sanibel Island showing the location of Blind Pass (right side of image, see arrow) where the inlet separated Sanibel and Captiva islands in 1958. *Betty Anholt*

The Captiva beach has a long history of erosion. Although much of it has been attributed to the 1921 opening of Redfish Pass and its subsequent enlargement by later storms, the erosion began well before that. Since 1903 (S. Bryant, pers. comm.) approximately 172 m (564 ft) has disappeared from the width of the island.

Steadily encroaching water removed much of the gulf-front Captiva Road by the mid-1950s and by 1963 approximately 1 kilometer of the paved road had been swept away and was vacated by Lee County. Hoping to slow and control the erosion of this beach the Captiva Erosion Prevention District (CEPD) was formed by a Special Act of the Florida legislature at the behest of Captiva Island residents. One of the first actions of the district after its 1959 legislative authorization was systematically to place 134 zigzag assemblies of interlocked concrete structures, known as Budd Wall concrete "dog-bone" groins, on the severely eroded midsection of the Captiva gulf beach in 1961. These had little effect on slowing long-term erosion. In the late 1980s, periodic major hydraulic dredging would be used for future renourishment of the Captiva Island beach.

In the late 1960s Lee County entered into a cooperative agreement with USFWS, and tens of thousands of cubic yards of fill were removed from the Bailey Tract (BT) on Sanibel and hauled by truck to Captiva to stabilize the road.

In the years Florida was a U.S. Territory prior to attaining statehood in 1845, Sanibel Island was known as Sanybel Island. This is an English corruption of what is considered to be the original charted Spanish place name, Puerto Sur Nivel. Common colloquial use of

Captiva Island beach erosion at its worst, circa 1970. A set of the dog-bone groins, installed in hopes of slowing erosion, are in the center of the image. *Deb McQuade Gleason*

the name Sanybel in both dialect and the literature was expiring by the time Florida became a state. This spelling was most recently used in 1896 (Cushing 1896). The island has been known simply as Sanibel since 1888 when land was released from the public domain for homesteading. The U.S. Post Office identified the island's postmark as "Sanibel" circa 1895. Following the island's incorporation as a city in 1974, formal action by the first city council officially created the City of Sanibel.

CLIMATE

South Florida, the region in which Sanibel and Captiva islands are situated, is located in the global subtropical zone. The islands are situated about 315 km (195 mi) north of the imaginary line called the Tropic of Cancer. The strongest of winter weather cold fronts infrequently reduce air temperatures on the islands to below freezing. When this happens the winter temperatures may drop to a minimum -3°C (26°F). There are a number of published documents that pertain to the incursion of such temperatures into the region. In general, these events are short-lived but on even rarer occasions the cold snap may be prolonged and temperatures remain below freezing for several days (LeBuff 1998). Prolonged periods of harsh cold temperatures affect cold-intolerant tropical organisms decisively. With a cold snap the cultivated and naturalized tropical vegetation in the region suffers. Many cold-susceptible tropical trees and plants are frost-burned or are outright frozen and die. In December 1962 an estimated 75 percent of Sanibel's Jamaica Tall coconut palms (*Cocos nucifera*) were destroyed by a series of severe minimum temperatures. Many species of tropical fishes, amphibians, and reptiles that exist in southern Florida are also killed during such weather events. Cold winter weather is the key factor that establishes the lower peninsula as the northern periphery of the geographical distribution of many forms of tropical fishes, herpetofauna, and even marine mammals (e.g., common snook [*Centropomus undecimalis*], American crocodile [*Crocodylus acutus*], and Florida manatee[4] [*Trichechus manatus latirostris*]). Abundance of these species have historically fluctuated because of a variety of factors, including the effects of unusual cold temperatures.

During the summer and early fall, high levels of humidity and scorching temperatures attest to the tropical nature of the Sanibel and Captiva climate. During the summer months, maximum air temperatures can exceed 40°C (104°F) and gulf water temperatures can reach 32°C (89.6°F). Water temperature and humidity highs support tropical weather systems; the region has a long history of major hurricanes. Two significant hurricanes impacted the islands during the period under discussion. Hurricane Donna of 1960 and Hurricane Charley of 2004 are two memorable tropical cyclones that ravaged the islands, 44 years between strikes.

Storm-surge flooding is commonplace during a hurricane's passage, and the hurricane-caused flood of record occurred on October 15, 1873. On that date a 4.26 m (14 ft) storm surge was recorded at the Port of Punta Rassa. This small but once-bustling seaport, now a resort community across San Carlos Bay and to the north of Sanibel Island's Point Ybel, is situated near the gateway to the barrier islands—the Sanibel Causeway toll plaza.

Precipitation will be discussed under the Hydrology and Water Management section.

LAND USE HISTORY

Soon after their formation both Sanibel and Captiva islands were used by aboriginal peoples. The most recent of these native groups were the now-extinct Calusa, which occupied the islands between 500 AD and the late 1700s until European-borne diseases and deadly conflicts with the Spanish invaders reduced their population. When the English took possession of Florida the first time, in 1763, the small remnant Calusa population, which had become Christianized, opted to migrate to Cuba with the Spanish. The Calusa have since been assimilated into the Cuban population.

Prior to the European incursions into Florida in the 16th century, the non-agrarian Calusa lived in harmony with their environment in a land of plenty. Today, their existence is known only from prehistoric habitations and the piles of refuse discarded from the seafood that sustained them. Their middens are simple heaps of shellfish, often extremely large, that once surrounded the Calusa's scattered waterside settlements. Archaeologists consider the center of their political and religious culture to have been located a few kilometers south of Sanibel Island, on Mound Key in Estero Bay.

By 1831, a group of investors in New York organized the Florida Land Company and acquired title to Sanybel Island, then part of a huge Spanish Land Grant that included most of Central Florida. By 1833, this consortium had a plat map of its proposed development engineered and was offering land for sale. The chief attraction for real estate on Sanybel Island was not its wonderful shell-strewn beach, but land for farming. Most of the island was subdivided into large bay-to-gulf parcels, but a town site on the island's eastern section, including roads and parks, was part of the design. The island was never permanently settled—the horrific mosquito population had a lot to do with that. The U.S. Army also openly discouraged settlement of the barrier islands during the Second Seminole War (1835 to 1842).

The earliest serious agricultural attempt on the islands occurred in 1869 when a farmer named William Allen attempted to cultivate castor beans commercially on Sanibel. This proved unsuccessful (Anholt 1998), but castor bean trees still germinate whenever upland is disturbed on certain parts of the island. The seeds from which these plants continue to germinate originated with the Allen crop of 1869.

In 1883, after the decision was made to establish the Sanibel Island Light Station (SILS) on the island's eastern tip (Point Ybel), all land on Sanibel Island was withdrawn and included in the SILS Reservation. The lighthouse became operational in August 1884, and the first permanent residents of Sanibel Island were lightkeepers and their dependents who moved into the new lighthouse quarters. Four years later the U.S. Lighthouse Establishment determined that most of the island to the west of the lighthouse was surplus to their needs. All land to the west and north of what is now Bailey Road was released for homesteading. This land was soon granted to applicants and as required, the arable land was cleared and eventually cultivated in truck crops. This produce included fine island-grown tomatoes; marketed as "Sanibel tomatoes," they became very popular in the northern markets. Squash, eggplants, and peppers were grown on the land adjacent to the Sanibel Slough; the higher, better-drained sections of land produced a variety of citrus.

This farming economy was successful, and despite the hardships of periodic drought, voracious clouds of mosquitoes, and flooding tropical storms and hurricanes, the farmers held on and many prospered—that is, until 1926. In September of that year a violent hurricane's storm surge inundated Sanibel Island and the prolonged flooding by seawater temporarily ruined the once-productive soil. As a result the farming industry ended on Sanibel and Captiva islands and never recovered. The abandoned, fallow farm fields disappeared in short order as native vegetation reclaimed the land. In certain locations on Sanibel the furrows made by mule-drawn plows are still visible on the ground's surface. The location of once-productive farmlands on Sanibel prior to the 1926 hurricane is depicted on Map 2.

Map 2. Formerly cultivated areas of Sanibel Island. Adapted from Cooley (1955). *SCCF*

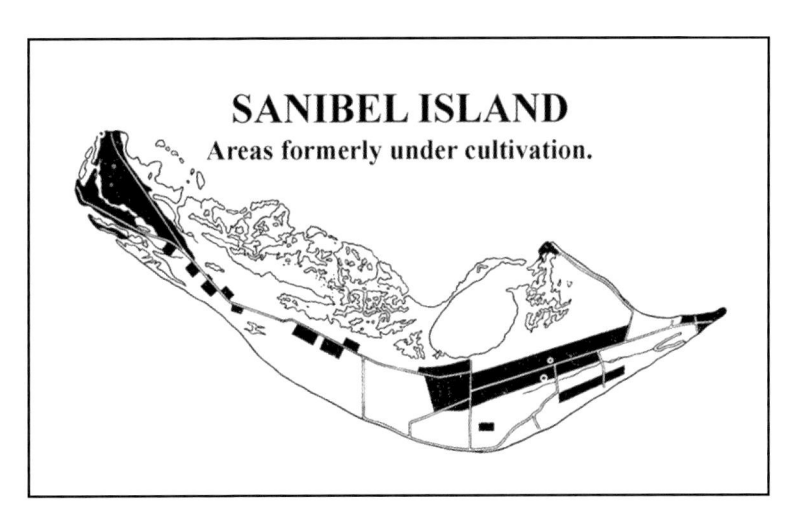

Today, land use on Sanibel Island complies with a Comprehensive Land Use Plan that was adopted by the COS in 1976. This document has been amended through time, but remains substantially as originally conceived. Major parcels of conserved lands on Sanibel Island are owned and leased by the federal government's USFWS and administered as JNDDNWR (2,630 hectares [ha]). SCCF owns and manages another grouping of tracts totaling more than 526 ha. Lee County and COS own a combined total of more than 81 ha. Much of this county- and city-owned land is dedicated park land, but the city designates most of its real estate as environmentally sensitive land.

In 1939, a Special Act of the Florida legislature (House Bill 1095) provided that all of Sanibel and a small section of southern Captiva Island were included in a new state wildlife refuge to be administered by the Florida Game and Freshwater Fish Commission (FGFWFC). An organization of interested islanders, the Inter-Island Conservation Association (IICA), had been established earlier and worked feverishly for wildlife protection on Sanibel and Captiva islands. The Special Act of 1939 also established the Lee County Refuge Commission, an appointed group of island citizens who would administer the state wildlife refuge.

IICA continued its efforts until SNWR was created on December 1, 1945, when a lease document was executed between USFWS and the Florida Trustees of the Internal Improvement Trust Fund and the Florida School Board.

SNWR was established during the administration of President Harry S. Truman. The National Wildlife Refuge System is part of the U.S. Fish and Wildlife Service of the

Department of the Interior. Later, with authority of Presidential Proclamation 2758, dated December 3, 1947, a Closure Order was published in the *Federal Register*. This action closed lands and waters within a legally described boundary to hunting, killing, or taking of migratory birds. The boundary includes all upland and wetlands that make up Sanibel Island; a small portion of southern Captiva Island was also included in the boundary as were some waters of the immediate estuary.

HYDROLOGY AND WATER MANAGEMENT

Water quality around the barrier islands has become degraded over recent decades. Contaminated discharges of water from Lake Okeechobee, which join nutrient-laden water from the Caloosahatchee watershed, impact the estuary and outlying islands. This contaminated water has affected important marine resources adversely and has been the subject of much controversy (Solecki et al. 1999).

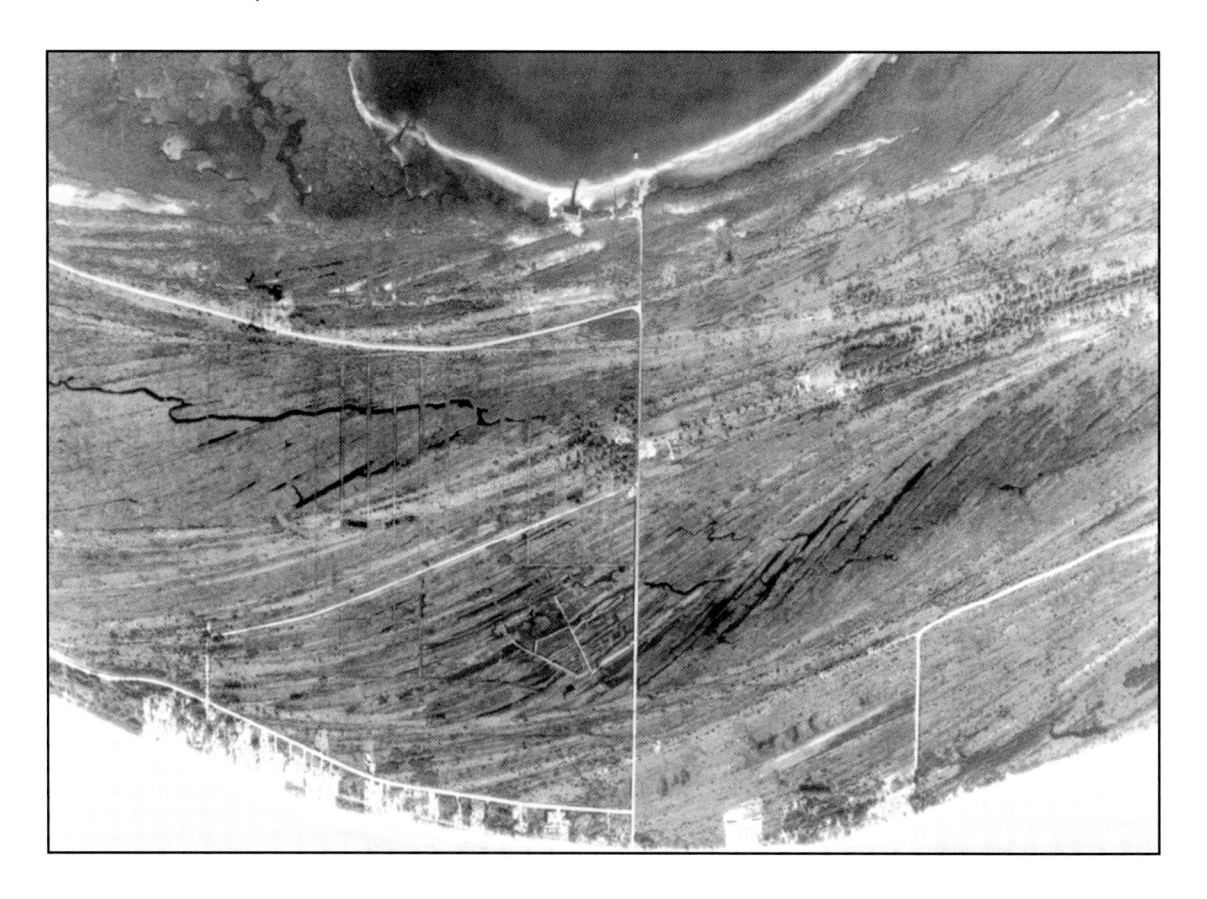

Central Sanibel Island, circa 1953, prior to beginning of the ditching program of LCMCD. Many of the ridge sets and swales are clearly visible in this photograph. The north to south road in the center is Tarpon Bay Road. The triangular dikes to its left are from early wildlife-management activities in the Bailey Tract. *James Pickens*

An unusual ancient system of paralleling upland shell ridges and depressed wetland swales runs between the gulf beach and the mangrove forest on Sanibel Island. Missimer (1973) identified seven distinct major ridge sets and scores of subsets that merged and formed present-day Sanibel Island.

Two of the major ridge sets are generally represented by the Gulf Beach and Mid-Island Ridge zones. These are discussed in detail beginning on page 12. The lower and often noncontiguous ridge subsets are the numerous ancient storm-created ridges that are situated between the aforementioned major ridges and randomly positioned in the expanse of the interior wetland basin. The swales historically flooded during common summer precipitation events. This wetland area is traditionally known as the Sanibel Slough (LeBuff 1998).

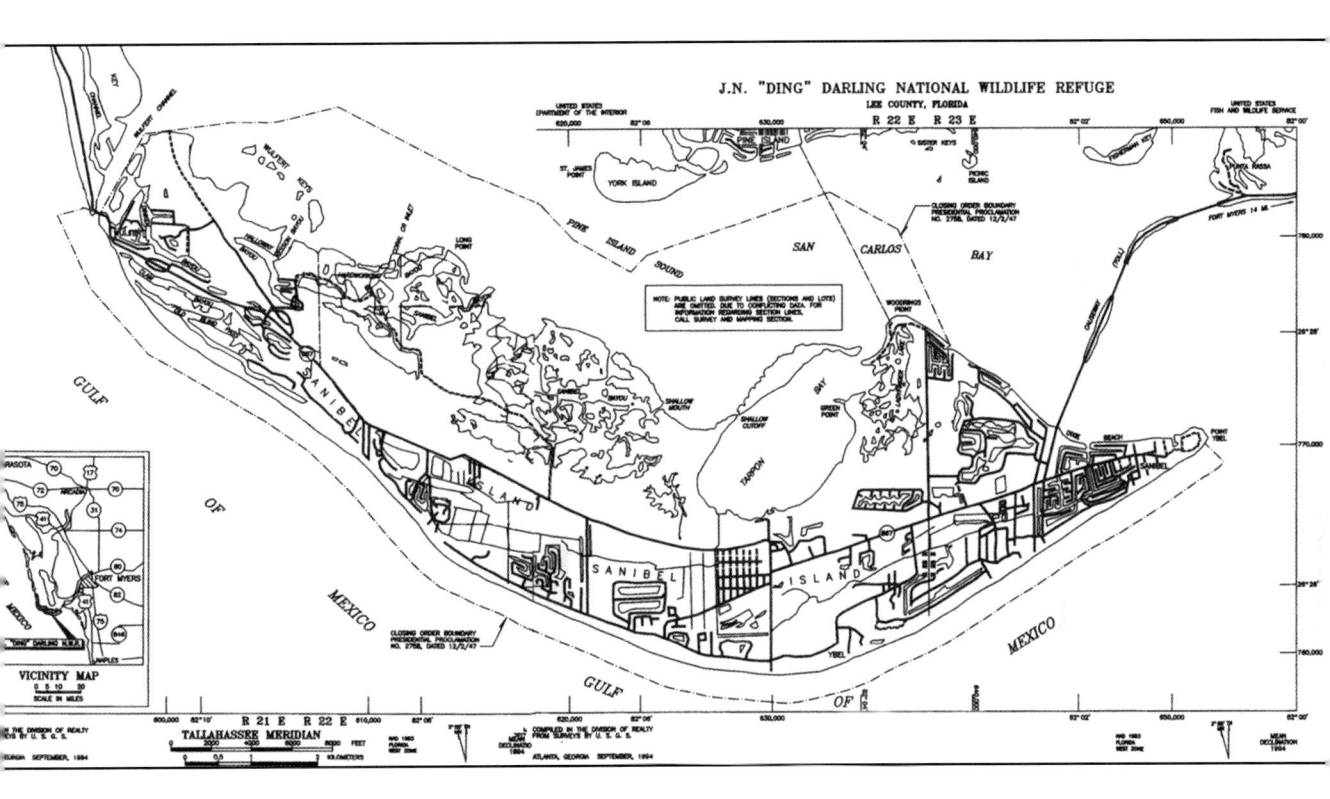

Map 3. Sanibel Island, Lee County, Florida. This map is modified from the two-map 1994 status map set for the J. N. "Ding" Darling National Wildlife Refuge. The dashed line around the island delineates the perimeter of the closing order boundary of 1947. *USFWS*

The region has an annual well-defined rainy season, usually beginning in mid-June and continuing until late September. Driven by massive thunderstorms, a line of ominous thunderheads builds up over the interior of the Florida peninsula on a daily schedule. By the time the typical morning onshore sea breeze subsides, these storms rapidly drift westerly and reach the coast and offshore waters where they climax, dissipating over the Gulf of Mexico. Powerful squall-driven straight-line winds, severe lightning, and torrential rainfall sweep across the barrier islands. It is not uncommon for one of these large systems to generate multiple waterspouts, powerful wind downbursts, or drop as much as 130 mm (5.25 in) of rain per hour. Some microbursts generated by large thunderheads can produce very damaging winds that can attain a wind velocity in excess of 250 km per hour (km/h) (155 mi per hour [mph]). Lightning strikes during these events are very frequent and extremely hazardous; they have a history of causing significant property damage, igniting wildfires, and even causing human fatalities on the barrier islands.

Generally, rainfall on Sanibel Island is not distributed uniformly, and readings are variable from one end of the island to the other. A precipitation recording station in JNDDNWR was centrally located on the island and was regularly monitored for nearly 20 years until it was relocated in 2008. During the decade between 1990 and 1999 this gauge collected an annual average of 180 cm (71 in) of precipitation.

Another unique aspect of Sanibel Island's hydrology is the presence of an impermeable layer of clay at various depths beneath the island's surface (Boggess 1974). Prior to 1956, rainwater that was trapped and retained in this basin would seasonally evaporate, or self-drain laterally as seepage through the enclosing ridges. During very wet summers (e.g., 1949), once the unified surface water aquifer of the saturated wetlands reached a maximum elevation of 1.3208 m (52 in) above mean sea level (MSL), it broke through the frontal dunes of the Gulf Beach Ridge Zone at a customary site on eastern Sanibel Island, 1.3 km (0.8 mi) west of SILS. Because of this flowing drainage, early island residents named the main swale of the interior wetlands the Sanibel River, and the major canal that has meandered through the freshwater wetland system since mosquito ditching ended there in 1962 is known by that name today. For several weeks after the trapped freshwater breached the gulf beach ridge, nearly all of the freshwater would drain into the Gulf of Mexico. An estimated 7 million gallons of freshwater drained away annually from the Sanibel Slough into the Gulf of Mexico during this temporary breach (Provost 1953). The historical hydroperiod of the seasonal wetlands of Sanibel Island was a dramatically simple wet-dry cycle because of this drainage factor.

Historically, fewer than 20 established alligator holes were known in the Sanibel Slough before 1950 (W. Wood, pers. comm.). These holes were the sole providers of freshwater for migratory and resident wildlife species during the driest part of the year. In addition, there were nine scattered seasonally fresh, rainwater-flooded, natural open ponds, each of which hosted multiple alligator dens. Through the efforts of resident alligators, the entrance of each alligator cave remained flooded, even at the lowest of water levels. Prior to mosquito ditching, the critical role that American alligators played in this barrier island's ecology was essential to the survival of aquatic animals and those that preyed on them.

In 1954 USFWS purchased a 40.5-ha (100-acre [ac]) parcel from Frank P. Bailey in the heart of the Sanibel Slough. USFWS had previously leased the property from the Baileys, and it was named the Bailey Tract (BT). About a year later, when water levels were at their lowest, USFWS attempted some basic water-management practices in hopes of creating additional year-round, surface freshwater pools within the BT. A number of potholes were dynamited in the dry marsh around a colonial bird rookery popularly known as the Mangrove Head. This was a prominent island-like feature, a landlocked stand of red mangroves (*Rhi-*

zophora mangle) in the central section of the Sanibel Slough. With the next cycle of seasonal rains, the blasted depressions flooded, and over time alligators found and used them. The crocodilians continued to maintain the depressions as they would one of their self-excavated holes.

From 1954 to 1958, USFWS drilled three artesian wells in the BT, and J. N. "Ding" Darling personally financed the first of these. The hope was that the flowing wells would be a source of water for the marsh during the driest part of the year, but the water was not pumped and the artesian flow was insufficient to keep the nearby impoundment pools within the small dike systems wet. Groups of minnows and adult and larval anurans were concentrated at the well sites during drought periods in the late spring. Later, however, large spoil ponds were excavated in the BT, and after the wells were capped in the 1970s, they were no longer allowed to flow freely.

MOSQUITO CONTROL

A further discussion of mosquito-control activities is essential to better understanding the major habitat alterations that were part of Sanibel Island's recent history, which have impacted the island's herpetofauna over time. A record New Jersey light-trap collection of saltmarsh mosquitoes (*Aedes taeniorhynchus*) was made in 1950 on Sanibel Island (LeBuff 1998). The Sanibel-Captiva Mosquito Control District was formed in 1953, and islandwide mosquito control was initially attempted using the pesticide dichloro-diphenyl-trichloroethane (DDT) as its principal agent. Unable to cope with the mosquito problem, the district was abolished and mosquito abatement on the island became the responsibility of the Lee County Mosquito Control District (LCMCD), formed by the Florida legislature in early 1958. Early on, LCMCD continued the use of DDT as an adulticide, but by the latter part of 1958 Director T. Wayne Miller ended application of that organophosphate, and another, Malathion, became the pesticide of choice. This would be replaced with Fenthion in 1965. The possible impacts of these adulticides and the larvicides that were often used in conjunction with them on the herpetofauna of Sanibel Island, especially with regard to amphibians, is discussed later in the Account of Species.

Miller incorporated a concept recommended by the Florida State Board of Health (Provost 1953). LCMCD launched a vigorous mosquito-control program on Sanibel Island when the ditching program was launched in 1958 (LeBuff 1998). By the middle of the next decade, the entire wetland basin of the Sanibel Slough had been crisscrossed and channelized by many kilometers of interconnected ditches that physically reached each swale. These seasonally rainfall-flooded depressions had produced the world-class numbers of *Aedes taeniorhynchus*. By tying the surface water aquifer together when water levels in the rain-swollen wetlands peaked, mosquito larvae-eating fishes (e.g., *Gambusia*) were available and biological mosquito control came to Sanibel Island. Ditching created permanent surface water on Sanibel Island, and this program directly benefited some, but not all, of the island's herpetofauna. Habitat changed quickly and dramatically, and later increases in the number of borrow ponds for real estate development would significantly increase the acre-feet of available surface water for all resident wildlife species and migratory birds.

ISLAND ECOSYSTEMS

The first Sanibel Island vegetative map was developed by George Cooley in 1955 (Map 4). He identified seven major plant habitats that form the basic cross-section of the island. Twenty-one years later, in his *Sanibel Report* (1976), John Clark adopted Cooley's principal major ecosystems and identified them as:

- A. **Gulf beach**
- B. **Gulf beach ridge**
- C. **Interior wetland basin**
- D. **Mid-island ridge**
- E. **Mangroves**
- F. **Tidal flats**
- G. **Bay beach**

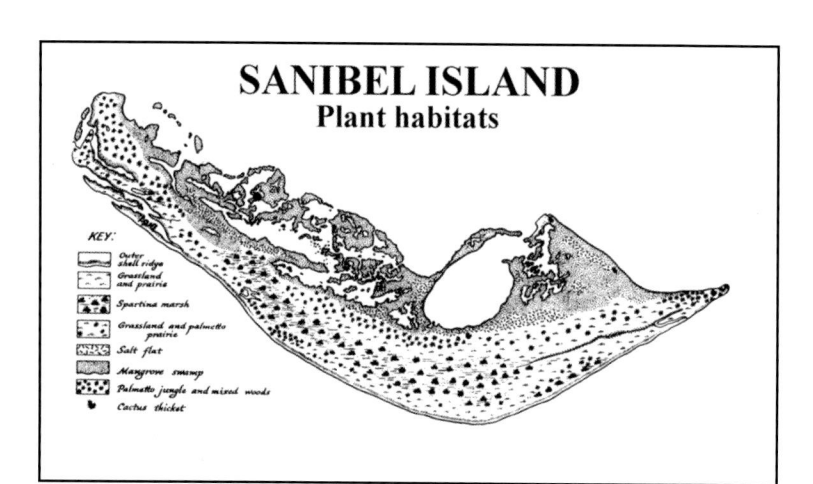

Map 4. The principal habitats and vegetation associations on Sanibel Island. Adapted from Cooley (1955). *SCCF*

We opted to follow Clark's ecosystem definitions to define habitat preferences of amphibians and reptiles, and these will be used later in the individual species accounts. More recently, these distinctive habitat types have been redefined by COS and are now designated as zones which are basically explained by their respective elevations, vegetative types, or water regimens. Their description and each habitat's specific relationship to the ecological niches occupied by the herpetofauna of Sanibel Island are outlined here.

A. GULF BEACH ZONE

Sanibel and Captiva islands' open frontal beaches and their integral dune system typify the Gulf Beach Zone and are contiguous along the islands' Gulf of Mexico shoreline. In 1971 the State of Florida established a Coastal Construction Control Line (CCCL), which included the sandy beaches of Lee County's barrier islands. This variable demarcation is based on the relationship of wave energy and beach profile at predetermined transect points. It is positioned at a variable setback, roughly 50 m (164 ft) upland of the established mean high water (MHW) line. We consider the Gulf Beach Zone to be totally within the CCCL. According to Florida law, the beach of the Gulf Beach Zone that is below the MHW line is public beach and not subject to private property rights that would otherwise deny public use or passage.

The unspoiled gulf beach of Sanibel Island is uncrowded in the residential area along West Gulf Drive. COS policy leaves the beach in its natural state, and city ordinances prohibit mechanical cleaning. By prohibiting destructive activities, sea turtle and shorebird nests are protected. *Charles LeBuff*

In 1959 the Gulf Beach Zones of Sanibel and Captiva islands were separated by a functioning inlet known as Blind Pass. At the time, this almost river-like narrow opening into the Gulf of Mexico was located 4.22 km (2.6 mi) south of the position of the Blind Pass Bridge of modern times (see photograph on page 3). The course of this changeable pass, notwithstanding the presence of Clam Bayou, was neither wide nor deep, and its flow capacity never moved volumes of water through it as is typical of other area passes, such as Redfish and Captiva passes. In fact, it was an opening of minor significance when it came to interchange of tidal waters. After 1959 the Blind Pass opening migrated north and completely closed by 1967. Sanibel and Captiva islands were then joined together by a common Gulf Beach Zone.

Blind Pass has naturally reopened and closed several times since then. As of 2012, it remains open after a second major dredging project followed one in 2009 that cost $3.2 million. Its capacity to remain functional is questionable.

There is a perception among some government officials and environmental groups that an open Blind Pass is essential to the vitality of the Pine Island Sound ecosystem, but the repeated opening and closing of the pass makes this seem doubtful.

Prior to designation of this CCCL and the birth of COS in 1974, Lee County established a bulkhead line that defined a construction setback line for coastal development. This allowed placement of concrete sheet pilings, or seawalls, at a distance of 14.25 m (50 ft) landward of the mean high tide line along the gulf beach and a shorter distance along the bay beach of the barrier islands. Some pre-COS resort housing developments along the Gulf Beach Zone were built using this form of beach armoring. This rule was ostensibly passed to protect private property from erosion. Even then there was considerable public sentiment on Sanibel and Captiva islands to prevent adoption of such a policy, but, as usual, the county didn't listen to the environmentally concerned island residents and voters.

This became another talking point for those islanders who were part of the groundswell that was building for incorporation. Most residents of Sanibel Island viewed the option of breaking away from the seeming incompetence and regulatory stumbling of the county commissioners at the time as the only way to protect the future of their island and their quality of life. When it came to a vote on the matter in November 1974, Sanibel's voters overwhelmingly opted for incorporation, and the island city was established.

The Gulf Beach Zone is vegetated with hardy plant species that are capable of withstanding the poor soil conditions, frequent high winds, and intermittent salt water overwash. Typical plant species commonly found in this zone include: railroad vine (*Ipomoea pes-caprae*), sea purslane (*Sesuvium portulacastrum*), sea oats (*Uniola paniculata*), nickerbean (*Caesalpinia bonduc*), and bay cedar (*Suriana maritima*).

The Gulf Beach Zone provides important foraging and nesting habitat for a variety of ground-nesting shore- and waterbirds such as plovers and terns. The open, usually unvegetated dry beach also provides prime nesting habitat for four species of sea turtles: the loggerhead (*Caretta caretta*), green (*Chelonia mydas*), leatherback (*Dermochelys coriacea*), and Kemp's ridley (*Lepidochelys kempii*), as well as the smaller ornate diamond-backed terrapin (*Malaclemys terrapin macrospilota*).

B. GULF BEACH RIDGE ZONE

In some descriptive accounts of beach-related ecosystems, this zone is often classified as the scrub transition zone. On Sanibel and Captiva islands the terrain of this dry, well-drained zone contains the highest natural elevations to be found on the islands (about 4 m [13 ft] above MSL). In the case of Sanibel Island this zone separates the Gulf Beach Zone from the Interior Wetland Basin Zone.

Herpetofauna utilizing this ecological zone are typically those that are not entirely moisture-dependent for habitat or prey. This zone has historically supported such species as the southern toad (*Anaxyrus terrestris*); six-lined racerunner (*Aspidoscelis sexlineata*), eastern diamond-backed rattlesnake (*Crotalus adamanteus*), eastern indigo (*Drymarchon couperi*), and coachwhip (*Coluber f. flagellum*) snakes; Florida box turtle (*Terrapene carolina bauri*); and gopher tortoise (*Gopherus polyphemus*). The marginal or transition area between

This section of the Gulf Beach Ridge Zone along West Gulf Drive, Sanibel Island, is typical of this ecosystem. This photo represents the area having the highest natural elevation on the island. *Charles LeBuff*

this zone and the Gulf Beach Zone also provides nesting habitat for sea turtles. The Gulf Beach Zone vegetation listed above extends onto the waterside toe of this system, but the usually dense canopy in the now rare undisturbed systems include joewood (*Jacquinia keyensis*), seagrape (*Coccoloba uvifera*), gumbo limbo (*Bursera simaruba*), and cabbage palm (*Sabal palmetto*). Many of the amphibians and reptiles known to occur on the islands occupy some part of this zone during their life cycle.

C. INTERIOR WETLAND BASIN ZONE

These nontidal wetlands are generally situated between the Gulf Beach and Mid-Island Ridge zones. They are a series of usually dry, former oceanfront dune ridges of low elevation and seasonally precipitation-flooded swales that separate the upland ridges from their predecessor ridge set. The swales are indicative of subsequent storm events that lacked the power or directional configuration to connect a new storm-generated dune with that of an older dune already in place. These interior ridge sets were historically subjected to periodic

A section of the modern Sanibel Slough in the BT is pictured above. Although considerably different from its historical vegetative composition, such *Spartina*-dominated wetland-lowlands still exist on Sanibel Island. Curtailment of seasonal flood-stage water elevations in the whole of this Interior Wetland Basin Zone has changed its vegetative profile. In the background is a strand of 25-year-old native buttonwood trees (*Conocarpus erectus*) that vigorously invaded this habitat when the noxious invader, Brazilian pepper (*Schinus terebinthifolius*), was removed. If the goal of land managers is to return the Sanibel Slough to its predevelopment condition, then temporary spikes in freshwater elevations must be restored. This would require operating a series of expensive dikes and pumping stations to drown out woody vegetation seasonally, followed by regular prescribed burning. *Charles LeBuff*

flooding. Together these lowland ridges and swales form a large basin-like interior wetland system. This basin, encircled by the gulf beach and mid-island ridges, contains approximately 1,486 ha of the combined lowland wetlands. This basin is known as the Sanibel Slough, and its unique hydroperiods are discussed elsewhere.

    The shallower swales and the margins of the deepest depressions of the interior wetlands of Sanibel Island, the latter being seasonal ponds that once contained a limited amount of exposed open water, were historically vegetated by cordgrass (*Spartina bakeri*) and sawgrass (*Cladium jamaicense*). By 1955, cattails (*Typha latifolia*) had invaded those larger ponds and reduced their viability as wildlife habitat. Red mangroves (*Rhizophora mangle*) and buttonwood (*Conocarpus erectus*) also germinated in the deeper sections of the slough. The most prominent of these water bodies were Stewart Pond at the western section of the slough and the Mangrove Head of the BT in the slough's central region.

Mangroves had originally entered the interior wetlands as propagules, transported simultaneously with storm surge, and some may have entered the system during those periods when the flooded slough drained into the Gulf of Mexico. Fire and flood-stage water levels contributed to the healthy biological integrity of this unique wetland system. Over time it remained balanced, but because of a dearth of major hurricanes and coincidental saltwater flooding in the latter half of the 20th century, along with demands for a lowered maximum water level by property owners whose home sites were seasonally flooded, the system was disrupted. Coupled with a change in the natural hydroperiods were the island community's understandable but environmentally foolish fire-control practices. The ecological health of the system suffered when wildfires, an important factor that helped make the Interior Wetland Zone a healthy and successful ecosystem, were extinguished immediately and not allowed to burn through the Sanibel Slough.

Extensive and widespread mechanical pest-plant control operations in the Interior Wetlands and Mid-Island Ridge zones in the latter third of the 20th century were designed to eliminate the exotic Brazilian pepper (*Schinus terebinthifolius*). Unfortunately, this technique only facilitated and accelerated significant plant transitions in those zones. The monotypic stands of Brazilian pepper were replaced by monotypic stands of the native buttonwood. As a result, the sought-after habitat improvement conceived and envisioned by land managers has not occurred. This is primarily because of the less-than-desirable maximum water levels that have been applied in attempted management of the wetlands. Throughout the system, reduced water levels have contributed to degradation as the slough ecosystem has drifted further away from its historical role as a productive and self-sustaining wildlife habitat.

Many amphibians, snakes, turtles, and the American alligator occupy and thrive in this important ecozone.

D. MID-ISLAND RIDGE ZONE

The Mid-Island Ridge Zone basically separates the interior wetland zone from the tidal mangrove zone. For all intents and purposes most of Sanibel-Captiva Road follows this central ancient ridge. Geologically the mid-island ridge developed into two major ridge sets that diverge and separate in the central section of Sanibel Island. One ridge supports the alignment of the inter-island road, whereas the other ridge angles from it in a more northerly direction and lies completely within JNDDNWR. This northern ridge is the most ancient on Sanibel Island. Isolated from the dynamics of its former Gulf Beach Zone origins for thousands of years, this ridge contains the only natural stand of live oaks (*Quercus virginiana*) on Sanibel Island.

One of the few remaining natural open meadows of Sanibel Island's Mid-Island Ridge Zone in 2012. This area on JNDDNWR once supported a colony of gopher tortoises (*Gopherus polyphemus*), and in those days the coarse grasses were kept well cropped by the turtles. *Charles LeBuff*

In the wetland triangle that formed because of the offset alignment of these two ridge sets, a section of the Sanibel Slough is preserved in near-pristine condition. This wetland triangle has escaped major ground surface alteration, other than a few acres at its eastern end that were cleared for a proposed commercial development. The project was never completed beyond land clearing that occurred just before the refuge acquired the impacted land. Since then, minor alterations have included the mechanical clearing of pest plants and more recently the construction of fire lanes by the refuge. It is noteworthy that this slough extension was never ditched for mosquito-control purposes.

Historically, the herpetological fauna utilizing the habitat of the Mid-Island Ridge Zone was comparable in species composition to that of the Gulf Beach Ridge Zone. Over time, however, it has been drastically altered as invasive exotic plants have crowded out important open-meadow type habitat. This latter habitat, once heavily occupied by upland species such as gopher tortoises and indigo snakes, has almost disappeared from Sanibel Island.

E. MANGROVE ZONE

The Mangrove Zone extends northerly from the narrow corridor of Sanibel-Captiva Road and the central mid-island ridge it follows, to the waters of Tarpon Bay and Pine Island Sound. Primarily vegetated by red mangrove (*Rhizophora mangle*), this zone consists of shallow internal tidal flats, lakes, open bayous, and open and tunnel-like creeks. Other mangroves, the black (*Avicennia germinans*) and the white (*Laguncularia racemosa*), occur in this system but are usually found at the tidal toe of the uplands in the intertidal zone, or in naturally formed depressions—basins situated within the vast red mangrove forest. These pools are typically vegetated with black mangroves and are infrequently flooded by wind-driven spring tides; more often, however, they are flooded by storm surges and precipitation. Even short-term flooding by seawater during storm events within these small natural impoundments will cover the pneumatophores of the trees and soon kill the black mangroves. Salt condensation coupled with flooding caused black mangrove mortality in the region post-Hurricane Donna in 1960.

The uplands situated within the Mangrove Zone of Sanibel may be manmade (e.g., Calusa middens, Wildlife Drive, Dixie Beach Boulevard) or small naturally created ancient ridges hidden within the zone.

Map 5. The CHES is expansive and includes the Mangrove Zone of Sanibel Island and the similar but smaller system on Captiva Island. This boundary of the CHES is located in approximation above and is the water areas delineated within the dashed lines.

The tidal mangrove forest is a key element in the health of the estuary and the system's overall productivity. Leaf fall, destined for waterborne dispersal from the mangrove forest into the estuary, provides valuable vegetative detritus that through a unique breakdown process contributes directly to the health and productivity of the marine environment and the value of marine resources. It is because of this productivity that the red mangrove system has long been recognized as an integral part of a complex nursery for the larval development of valuable commercial and sport fisheries. The mangrove forest also provides foraging and nesting habitat for a variety of colonial nesting birds.

Amphibians and reptiles that regularly range into and occupy parts of the mangrove forest include squirrel and Cuban tree frogs (*Hyla squirella* and *Osteopilus septentrionalis*), green anole (*Anolis carolinensis*), American alligator (*Alligator mississippiensis*), and the eastern ratsnake (*Pantherophis alleghaniensis*). Other species that selectively favor the mangrove forest as their primary habitat include the American crocodile (*Crocodylus acutus*), ornate diamond-backed terrapin (*Malaclemys terrapin macrospilota*), and mangrove saltmarsh snake (*Nerodia clarkii compressicauda*).

The mangrove forests of these barrier islands are dominated by the red mangrove (*Rhizophora mangle*), which is easily identified by its unique prop roots. Black (*Avicennia germinans*) and white (*Laguncularia racemosa*) mangroves are widespread through the system but occupy higher elevations than the red mangrove. *Charles LeBuff*

F. TIDAL FLATS ZONE

The Tidal Flats Zone is essentially the intertidal zone of the red mangrove forest. These shallow areas are closely associated with the Mangrove Zone, and much of the water bottom is intermittently exposed during the tidal cycles. Only the deeper channels, creeks, and bayous in the Tidal Flats Zone remain flooded, even during the most negative neap tides or weather-induced extremely low tides. The tidal amplitudes range between 10 (low) and 60 (high) cm (4 to 24 in) for Sanibel and Captiva islands.

For a few hours on either side of low tide, hundreds of hectares of tidal flats are exposed inside JNDDNWR and other areas of Sanibel and Captiva islands. Crustaceans and other invertebrates dwelling in the then-accessible bottoms have always sustained large populations of wading birds. This image was taken in the winter when strong northerly winds influence the amplitude of the tidal event and create longer-lasting and lower tides. *Charles LeBuff*

The tidal rhythms of the coastal waters of the eastern Gulf of Mexico are characteristically unique in the Western Atlantic Ocean because of their inequality with standard tidal curves. Typically, tides around the Earth are placed in three categories and classified as: (1) *semi-diurnal*, having two high and two low tides per tidal day (these occur regularly on the east coast of Florida); (2) *diurnal*, having one high water and one low water attained during a tidal day (these occur along the northern Florida gulf coast to the west of Apalachicola); or

(3) *mixed tides*, which are often considered transitional between areas where either diurnal or semidiurnal tides are dominant (this tidal regimen is typical for the marine waters adjacent to Sanibel and Captiva islands. Mixed tides create an obvious diurnal inequality between the highest of high water and the lowest of low water. This tidal variance is often confusing, especially for someone coming to the region from an area where either diurnal or semidiurnal tides are the norm.

During peak spring tides, when the tidal height is at its greatest astronomical and nonstorm-produced elevation, the tidal flats are covered with water levels that flood well into the red mangrove system. When extreme neap tides occur, the tide is extremely low and the Sanibel Island tidal flats are exposed for a prolonged period. Even the tidal flats that extend offshore, those that are located beyond the mangrove periphery and extend a considerable distance into Pine Island Sound, are temporarily elevated above the water level. It is during normal low and neap tides that a variety of birds converge on the exposed parts of the tidal flat zone; low water transforms the flats into a complex and valuable foraging area for migratory and resident birds.

G. BAY BEACH ZONE

On Sanibel Island the Bay Beach Zone is a narrow noncontiguous ribbon of biogenic beach that extends along the water's edge of San Carlos Bay and Pine Island Sound. Approximately15.25 m (50 ft) wide[5], this zone is exposed to the waters of the bay and sound and follows the island's shoreline westward from Point Ybel to Woodring Point on Sanibel's eastern end. Today, the continuity of this beach is broken by one mangrove creek and narrowed elsewhere by mangroves—both natural components of this ecozone. At the western end of the Bay Beach Zone, near the foot of Woodring Point, there is a considerable length of beachfront where the mangrove zone encroaches to the water's edge and the dry beach becomes low, narrows, and nearly disappears.

Several manmade structures affect the integrity of this ecological zone and have, over time, destroyed much of a once remarkable and productive beach. The Sanibel Causeway lands on the bay beach and concrete sheet pilings are positioned along parts of the shoreline east of the causeway and along much of the west side of the bridge. Today, there is little dry bay beach left at the foot of these seawall structures at high tide because of the long-term erosion that has occurred since their installation.

Three manmade inlets now break through the bay beach between Point Ybel and the western end of Woodring Point. Two were present in 1959, but one of those has since been plugged. The two that were engineered since 1959 provide the residents of two separate bayfront subdivisions boat access to the bay.

Sanibel Island's Bay Beach Zone has been altered over time and today is somewhat discontinuous in sections. Narrow and relatively short, this zone is positioned along the shore of San Carlos Bay and Pine Island Sound. Historically, it extended from Point Ybel to Woodring Point, where it reaches the entrance of Tarpon Bay. Sections of this beach are occasionally important for nesting loggerhead sea turtles and more often for ornate diamond-backed terrapins. This section in the photograph is located immediately west of the Sanibel Causeway. *Charles LeBuff*

SPECIES RECRUITMENT

An interesting phenomenon of recruitment by mainland herpetofauna to Sanibel Island in the 20th century is important and worthy of mention. In the early years, when the senior author first started building a herpetofaunal database for Sanibel Island, it was common in late summer for floating masses of introduced water hyacinths (*Eichornia crassipes*) to literally bridge the bay and lower reaches of Pine Island Sound. This plant floated down the river carried by the current of the Caloosahatchee resulting from discharges from Lake Okeechobee by the U.S. Army Corps of Engineers; at times, hyacinths formed a solid shore-to-shore floating carpet. Many species undoubtedly used these mats to disperse across otherwise hostile saline waters to colonize the islands. Even when these plants were absent, some turtles, lizards, and snakes reached the bay either in the water or on floating debris and then made the crossing to Sanibel Island with the aid of downstream currents from the outflow of the Caloosahatchee. Many were surely swept out to sea by the turbulent ebb tides (Sheng 1996) they encountered during their accidental crossing attempt.

In the years before the causeway, observations of snakes of the genera *Pantherophis* and *Crotalus* swimming in San Carlos Bay were regularly reported to the refuge office by commercial and sports fishermen and ferry passengers. Pre-causeway (before 1963), most native representatives of the region's herpetofauna that made a safe passage and succeeded in reaching the ecosystems of Sanibel Island did so by water and under their own power and navigational abilities.

The water hyacinths were temporarily trapped between the river's outflow and flowing tides. The upper debris wrack lines of both the bay and gulf beaches of eastern Sanibel Island were covered with thick layers of the decomposing hyacinths in the late summer. Concurrently, the voluminous freshwater flow that swept these plants down the Caloosahatchee drastically reduced the salinity of San Carlos Bay. Salinity reached such low levels that it is conceivable that specimens of the above-mentioned genera could have survived the crossing during this window of limited opportunity. Once they arrived, however, there was no suitable entry or breach leading into Sanibel Island's interior freshwater wetland ecosystem through which the physiologically stressed aquatic amphibians could have successfully passed.

Over time, a limited number of anurans succeeded in reaching the freshwater wetlands of Sanibel Island after being transported by the floating bridge of vegetation or during radical salinity reductions as a result of torrential precipitation during severe weather events. Some, but not all, of these populations had been established on the island long before 1959. Other species common to the mainland that one might expect could colonize Sanibel Island apparently failed to do so. Some of those that did and survived to establish a beachhead were successful over time, although others seem to have survived for only a limited period.

Application of herbicides upstream by the Lee County Hyacinth Control District has since curtailed the massing of water hyacinths in the Caloosahatchee.

By May 1963, the Sanibel Island Causeway's system of bridges and spoil islands spanned San Carlos Bay and replaced the small automobile ferries. The causeway has since played a major role in the recruitment of herpetofauna to Sanibel and Captiva islands.

ENVIRONMENTAL AND OTHER THREATS TO THE ISLAND HERPETOFAUNA

Modification and degradation of aquatic habitat: Not only did the ditching program of LCMCD alter the interior wetlands of Sanibel Island, so did the development of subdivisions and commercial centers, which resulted in a loss of habitat for island wildlife. Investors wanted to reap the financial rewards presented to them following the opening of the causeway in 1963. The quality of life for island residents suffered because of this race for profits. Visitor loads soared, creating gridlock traffic on island roads, and land sales boomed as a result of people "discovering" Sanibel and Captiva islands.

After COS was incorporated to cope with the impacts caused by near-unrestrained land development and the new city government took control of Sanibel Island's destiny, subdivision expansion was tightly regulated beginning in 1974. (The people of Captiva opted out of joining their neighbors on Sanibel to become a unified city, and Captiva Island remains an unincorporated part of Lee County.) COS wisely regulated land development that uses cut-and-fill techniques to move spoil from a borrow site (a future real estate "lake") to contour adjacent land to buildable elevations. This method of building subdivisions in wetlands was standard practice in the Florida civil engineering and land development industries for years.

The early ditching program of LCMCD was different. The program consisted of shallow intrusion into the surface water aquifer. Very rarely did excavation with small draglines penetrate the impermeable layer beneath that aquifer to a depth sufficient to allow intrusion of saline water into the surface wetland habitat. In contrast, construction methods in the post-causeway subdivisions used heavy equipment and dewatering to excavate the borrow ponds. Contractors dug down to an extreme depth to yield sufficient fill to raise home-site elevations up to the engineer-designed grade. Borrow ponds at such island developments as Lake Murex, Little Lake Murex, East Rocks, The Dunes, and Gumbo Limbo were dug very deep, some in excess of 9.14 m (30 ft) *below sea level*[6]. Even the refuge's BT did not escape dewatering of deep spoil pits to allow easier removal of fill. The depths of the bottoms of four of the pools that now exist in the BT are excessive by today's standards. Their depths were reached after dewatering for removal of maximum volumes of fill dirt.

As mentioned earlier, this work was conducted through a cooperative agreement between the Lee County Board of County Commissioners and the USFWS. The Lee County Department of Transportation desperately needed a fill site on island to help stabilize Captiva Road because of severe beach erosion. Over several years, county-owned equipment or that of their contractors excavated spoil in the BT with self-propelled scraper pans, while water was simultaneously pumped from the excavation sites. This equipment moved fill to the northwest corner of the BT along Island Inn Road where it was stockpiled to dry for use as needed later on Captiva Island.

Saltwater intrusion occurred at all the above-mentioned sites because the earthen seal that separated the upper freshwater from the lower saline water was compromised. Today, the water column in these ponds is stratified—a freshwater layer results from precipitation consolidating near the surface, and a layer of dense saline water that has intruded from below is positioned near the bottom underneath the surface freshwater. During the summer, rainwater stratifies above the deeper saltwater because the salt-load weight (density) of the latter maintains a lower elevation than freshwater. During the driest months of the annual drought cycle (April and May), surface water evaporates and saltwater temporarily intrudes into the upper water column, reaching a higher elevation.

The high bottom salinity is probably a factor that limits turtle and fish use and prohibits their long-term survival in this manmade habitat. For example, the first Florida mud turtle collected by LeBuff was taken in April 1960 from the nearly dry Mangrove Head of the BT. That aquatic system was totally reconfigured and deepened by Lee County later in the 1960s, with the blessing of USFWS because the work was considered habitat enhancement. Today, the Mangrove Head and the adjacent deepened pools of the BT seem to be nearly turtle-free because of poor water quality resulting from saltwater intrusion. Over the past 50 years, the Florida mud turtle seems to have vacated this once-prime habitat.

Impacts from landscaping: After Hurricane Charley in 2004, re-landscaping of homes and commercial establishments was required on a large scale, and large numbers of ornamental plants and trees were transported to the islands from plant nurseries throughout parts of South Florida unaffected by the storm. The imported vegetation came from areas where many species of invasive non-native lizards were well established. The probability that this non-native herpetofauna was transported here was higher on Captiva, which suffered more hurricane damage and therefore received most of the vegetation. Eventually, the invasive colonizers may threaten native amphibians and reptiles.

Impacts of fire: The longtime application of prescribed fire to the several fire-dependent communities on Sanibel Island is invaluable to their ongoing success. Without fire, these ecosystems degraded as invasive plants and changes in the hydroperiods prevailed after 1974, and continue until the present.

Prior to 1926, some of the island's ridges and their margins were productive farmland. Since then, the composition of the vegetation has changed to second-growth succession upland, a mixed community with the native palm *Sabal palmetto* predominant. Fire ignites the cabbage palms and the frond leaf litter beneath them. The cabbage palm is a fire-tolerant species, but when its dead fronds are removed by fire, the habitat they provide is temporarily lost. The boots of fallen fronds that remain attached to the trunks of this palm are simultaneously burned away, and a section of prime habitat for several species is lost.

When a long ridge is burned, recolonization by the herpetofauna is a much lengthier process than in habitat that burns during a controlled burn in a flatwoods system. Post-burn, herpetofaunal recolonizers must migrate perpendicularly, then laterally, as inhospitable habitats force them to traverse convoluted pathways since there are no established populations directly available for migration from the lateral periphery of the burn site. Especially when flooded, Sanibel's wide wetland swales can present a barrier that prohibits recolonization other than from the fire boundary at each end of a burned ridge.

During burns on these ridges, mortality within the herpetofaunal community may be considerable, and it is reasonable to assume that populations are impacted accordingly. Some lizard and snake species may take years to recover to preburn densities.

Fire events in the areas dominated by the tropical hardwoods, which were traditional coralsnake-favored habitat on Sanibel Island, were almost nonexistent for nearly a century prior to the 1980s when serious wildlife habitat restoration efforts were launched and the general application of prescribed burns was seen as a key management tool. Prior to that, controlled burns were rarely undertaken, and when they were the burns were generally confined to the *Spartina*-dominated interior wetlands.

Decline and recovery of the eastern indigo snake: Eastern indigo snakes used to be prevalent on Sanibel and Captiva islands. Lechowicz has heard countless stories from residents who saw them frequently in their yards in the past and remarked at how beautiful these snakes were, but they have disappeared from the islands.

Eastern indigos have a very large home range. Males have been known to establish a territory as large as 5.4 square (sq) km (2.1 sq mi). The two largest conservation lands on these barrier islands straddle Sanibel-Captiva Road, and indigo snakes regularly traveled back and forth across this road on any given day. When you combine large snakes traveling long distances (across Sanibel-Captiva Road) foraging or looking for mates, daylight activity, prejudice against large snakes, and a large number of vehicles coincidental with the peak tourist season, it never leads to a good outcome. One by one, indigos were run over, with few, if any, new recruits reaching the islands from outside populations. Unfortunately, some people deliberately killed indigos, not knowing the species' rarity and protected status. Indigos also were occasionally collected by regional reptile hobbyists or for the national pet trade, but the major culprit in this snake's demise was traffic.

In the early 1990s, the JNDDNWR staff briefly considered reintroducing indigo snakes to Sanibel Island to help enhance the species on the island. LeBuff supported the concept but advised that the indigo snakes of Sanibel and Captiva islands were typically black-throated and did not have the red-orange-toned labials and gular areas typical in mainland populations.

Some nearby populations are colored similarly to those that occurred on Sanibel and Captiva islands in the past. Upper Captiva, Cayo Costa, and Pine Island still have indigo snake populations, but those on Pine Island are threatened because of high-volume traffic on Stringfellow Road (County Road 767). These populations are currently (2013) being studied by Lechowicz of SCCF in coordination with the Orianne Society[7]. Those on Upper Captiva and Cayo Costa are likely to survive long-term because they are not impacted by the nemesis of all native snakes—high-speed vehicular traffic.

The indigo snakes on the northern barrier islands of Lee County may have the same color variation that once was dominant in the Sanibel and Captiva islands populations and may be good candidates for repatriation to Sanibel Island. Repatriation of this species is a good concept, once habitat restoration provides a suitable series of ecosystems that will again support this species over the long term. A viable restructured population of the eastern indigo snake, however, will first require the development and adoption of a recovery plan. This will require funding and implementation of a set of unique and expensive conservation options if the managers of JNDDNWR and owners of other lands dedicated to habitat restoration and wildlife conservation are to be successful in such a lofty management goal.

A thoroughfare of major herpetofaunal mortality: Sanibel-Captiva Road presents the most serious threat to the herpetofauna of Sanibel Island. The eastern indigo snake is not the only form of native wildlife that has been seriously impacted by the traffic flowing on Sanibel-Captiva Road. Many other indigenous terrestrial, and some avian, wildlife species also suffer from losses as individuals, and in some cases family groups, attempt to cross this busy roadway.

Other than financial costs, the major obstacle to a successful indigo snake repatriation project is unequivocally linked to the threat of Sanibel-Captiva Road. The islands could have all the pristine, open, grassy, uplands similar to those that existed into the early 1970s and still could not support indigo snakes unless they were prevented from crossing this high-traffic-volume road.

There are two ways to accomplish this. One is to elevate the entire Sanibel-Captiva Road so that all wildlife could have unobstructed corridors to travel back and forth under the road. The other, less expensive, concept would require fencing off the entire perimeter of the road corridor at the extremes of its right of way, using a solid material or small-diameter fencing, and positioning open unobstructed eco-passes (box culverts or tunnels that go under the road) about every 0.4 km (0.25 mi) and at intersections along the length of the road, allowing animals to pass safely from one side to the other. Both solutions would be huge projects and by their nature very expensive, but it is not out of the realm of possibility for close-knit, conservation-oriented island communities such as Sanibel and Captiva. All forms of terrestrial wildlife on the islands would be direct beneficiaries, and the lands affected could be managed to their full potential for optimal wildlife values.

The need for wildlife corridors or eco-passes along Sanibel-Captiva Road is not a new concept. Roadkill surveys, conducted by members of the Sanibel Wildlife Committee[8] in the 1980s, showed distinct areas where a variety of wildlife species crossed roads in higher abundance than at other locations along the road network. Committee members supported the idea of helping wildlife cross safely at these high-traffic areas.

COS established precedence on Sanibel Island for such a public policy. Because of ongoing wildlife road kills, the city bowed to a boisterous level of public input and installed some poorly designed "critter crossings" under parts of Sanibel-Captiva Road. These were nothing more than standard concrete culverts without accessory wing walls to intercept and direct animals to the openings. The tubes were set at elevations that were far too low, and they remained flooded most of the year. As designed and implemented, this attempted quick fix proved to be basically worthless and a waste of the taxpayer's money.

Costs for a successful eco-pass fencing project could be underwritten by both public and private financing. The bridge-like underpass crossings would be engineered with a wide horizontal clearance that is spanned using a minimum number of supporting pilings and with a high-enough vertical clearance to allow passage of the tallest native mammals. They would, by design, be bridge-like, yet configured to prevent permanent flooding beneath them. This part of the cost could be borne by COS, and the overpasses could be budgeted and installed incrementally over time, keeping pace with the installation of the fence barriers.

Land acquisition for conservation purposes is basically complete on Sanibel Island. Each wildlife-related organization has its own complex, high-caliber institutional administrative and education center in place. Land management for habitat enhancement and wildlife diversity has started and now earmarked funds could be raised and grant monies sought by USFWS, SCCF, CROW, and COS. These organizations can begin the regimented installation of the eco-passes and protective and directional barriers along both sides of the Sanibel-Captiva Road corridor.

Many people would support such an effort to reduce the number of road-killed animals on the island, though others may consider it a waste of time and money. In fact, this work would be first-class land stewardship and wildlife conservation, something for which Sanibel Island is already noted. It could serve as a model for other communities and bring Sanibel international recognition for its responsible commitment to the quality of life of all its residents.

Aesthetics must be considered, of course. Unless design features are impeccable, some residents and visitors may consider a visible fence to be an eyesore that takes away from the beauty of Sanibel and Captiva islands. This can be avoided with an additional setback inside the property lines for a vegetative buffer—a green-screen—to help conceal the fences. There are efficient, functioning, similar-purpose fences in place at the National Key Deer Refuge on Big Pine Key in the Florida Keys and at the Florida Panther National Wildlife Refuge east of Naples, Florida. Sections of U.S. 41 are now being elevated where the highway passes through Everglades National Park.

EARLY HERPETOLOGICAL EVIDENCE

We have searched a range of mostly popular literature pertaining to the herpetology of Sanibel and Captiva islands and have uncovered little data that documents the composition or dynamics of the islands' herpetofaunal populations prior to 1959. Obscure newspaper

This vintage photograph appears on page 46 of *A Sanibel and Captiva Family Album.* According to the caption below it, this supposed rattlesnake was collected near Sanibel's Casa Ybel Resort, in 1911. In our opinion this is a kingsnake (*Lampropeltis*). *Island Reporter*

articles in the archives of the *Fort Myers News-Press* provided some news stories pertaining to island alligators or sea turtles, but little information exists that is useful for compiling even the most rudimentary early species list. In recent years, books that incorporate vintage photographs have provided some interesting evidence. In her excellent Sanibel history, *Sanibel's Story*, author/historian Betty Anholt gives us a glimpse into a few reptile species that may have occurred on the islands much earlier in the 20th century. Her work includes documentary evidence through the use of photographs that the alligator, gopher tortoise, loggerhead turtle, and eastern ratsnake occurred on Sanibel Island in the first half of the last century.

Another recent pictorial book contains early 20th century photographs that document the alligator, loggerhead sea turtle, and eastern indigo snake on Sanibel Island. There are two other photographs that if, as stated in the text, were actually taken on Sanibel are most unique. In the book, *A Sanibel and Captiva Family Album*, edited by Ralf Kircher and published by *Island Reporter* in 1996, there are two photographs that appear to portray an adult kingsnake (*Lampropeltis*), which the text misidentifies as a rattlesnake. The Florida kingsnake has never been recorded from Sanibel or Captiva islands anytime in the continuous 54-year period that record-keeping has been conducted. These snakes occur near the coast and on barrier islands in other parts of their range. In the specimen pictured here (from the abovementioned book), many of its features are difficult to ascertain, but the part of the specimen's posterior that is in focus clearly shows the typical kingsnake pattern. The pattern of this snake, however, varies somewhat from what we would expect to find on a specimen from the adjacent Florida mainland population of this species. The body pattern of the snake in this photograph is more typical of kingsnakes that occur much farther north in Florida. It is possible that this snake was caught elsewhere, or even was brought to the island as a captive . . . perhaps as someone's pet.

COLLECTION TECHNIQUES

In the beginning (1959), individuals of the resident herpetofauna of Sanibel and Captiva were collected coincidental to everyday refuge operations. Documented amphibians and reptiles were identified through collection and examination, and were not recorded based solely on casual observations or distant sightings. Periodically, habitats known to be occupied by the various indigenous species were routinely searched to evaluate the population levels.

It soon became public knowledge among island residents that a refuge staff member was cataloging Sanibel's amphibians and reptiles. During this period, the SNWR office was the sole island clearinghouse for all out-of-the-ordinary observations or collections of wildlife and marine organisms on both Sanibel and Captiva islands. No other entity received observations or reports relative to non-molluscan marine life or wildlife. Islanders routinely communicated with the refuge office whenever they discovered a large or unusual snake or turtle, or even a strange frog. This was also true when an unusual or rare bird species was observed on the islands, when an uncommon mammal was sighted or killed on island roads, or when unique fish species were caught. On the other hand, uncommon, even new species, of seashells were reported to the commercial shell shops where collection data was noted and the unusual specimens were usually purchased by the proprietors for their personal collections. In the winter season, many prominent conchologists and malacologists[9] visited the islands, and mollusks were often turned over to them for cataloging in museum collections and later publication in the scientific literature.

Whenever possible the senior author would respond and attempt to capture and relocate herpetological specimens, especially when the individual might be in danger as a result of human behavior. This was especially true for venomous snakes.

The roads of the islands were also inspected periodically by slowly driving a predetermined route at night. When both dead-on-road (DOR) and living amphibians and reptiles were discovered, they were collected for identification. This collection method was most productive during heavy rain when the Sanibel Slough was at flood stage and animals were forced to move to higher ground.

In 2002 a more systematic approach to sampling the herpetofauna of Sanibel and Captiva was implemented by the junior author. This was advantageous because in the intervening 50 years since herpetofaunal sampling was initiated the available habitat and the basic composition of species was altered, either by mosquito control operations or subdivision development. The stability of the amphibian and reptile population levels on the islands had dramatically and noticeably declined over the intervening half-century and required reassessment.

One of Chris Lechowicz's first tasks at SCCF was to develop an updated herpetofauna list. Since LeBuff's first compilation in the early 1960s several list upgrades have been made, among them were those of George Campbell, Steve Phillips at SCCF, and Lennie Jones, who was temporarily assigned to JNDDNWR. Lechowicz employed standard sampling methods that are used in other regions to accomplish his survey, including drift-fence sampling, frog-call surveys, aquatic turtle sampling (hoop and basking traps), roadkill surveys, and of course, public outreach.

The labor-intensive technique of drift-fence collecting was used systematically for the first time to sample various habitats for herpetofauna on Sanibel Island. Drift fences are barriers that divert a moving animal toward a trap. This works because most amphibians and reptiles do not attempt to climb over the barrier but rather follow the barricade, moving along it to the right or left. The barrier has either pitfall traps or funnel traps positioned strategically to capture the animals.

The drift-fence barriers used here were made from commercial silt fence or aluminum flashing. Pitfall traps were 19-liter (5-gallon) buckets buried with the opening at ground level. The funnel traps were homemade using aluminum screening or hardware cloth. Both types of traps were shaded to keep the animal safe when captured, and the buckets in the pitfall traps contained wet sponges to prevent desiccation in amphibians. This collection technique was used for snakes, lizards, some turtles, and frogs.

In 2002 Lechowicz initiated frog-call surveys sporadically through the summer and winter in different habitat types to record the presence of all Lee County, Florida, species. In 2006 he created a definitive frog-call route, which is now part of the regional Frog Watch Network. This survey is conducted every third Wednesday during the summer months in collaboration with JNDDNWR and SCCF.

Alligator surveys have been conducted on Sanibel Island with some regularity for more than 50 years, and occasional alligator attacks over the past 10 years have spawned more regular monitoring. Currently, these surveys are carried out in collaboration with JNDDNWR and SCCF. The counts are not meant to assess a total island population but rather to monitor trends. To estimate the actual number of alligators on the island, we would need to implement a mark/recapture study as LeBuff conducted in the 1960s. This technique is not currently favored by any of the three largest conservation land managers—JNDDNWR, SCCF, and COS because of liability. If someone were attacked by a marked animal, questions could be raised as to why this "dangerous" animal was not removed previously. Day and nighttime surveys are conducted two times a year on conservation lands and the three island golf courses.

Sampling the Sanibel Slough for aquatic turtles was an important step in assessing the diversity of the herpetofauna of Sanibel. The existing species lists had an extensive turtle section, and it was necessary to verify the presence of all previously listed species. As of 2013, all the turtles listed have been found in the river with the exception of the Florida mud turtle (*Kinosternon subrubrum steindachneri*), but the species was found elsewhere on Sanibel Island. This turtle has been reported as declining throughout the Florida peninsula, although the reason is unknown. Floating and submerged basking traps were used to capture basking turtles, as were partially submerged, baited hoop traps to catch the more aquatic non-herbivorous species.

Terrestrial and semi-terrestrial turtles such as the gopher tortoise and Florida box turtle are monitored in different ways. Gopher tortoise burrows are surveyed once a year in October and are classified as active, inactive, or abandoned on SCCF properties. JNDDNWR does this on their lands at around the same time. Many small neighborhood communities on Sanibel Island have started to monitor their populations using the same method. The stability of the island's population of Florida box turtles is currently at a high level of concern. At the time of this writing, an island-wide, mark-recapture study is being conducted by Lechowicz to monitor their populations.

Roadkill surveys are a very important part of herpetofauna sampling. Unfortunately, many amphibians and reptiles perish on our roads daily, and the data provided by seeking out these fatalities is extensive. Collecting roadkill data is incidental and not intentional, meaning time is not expended solely looking for roadkills, but rather the data is recorded in concert with other operations. Some of the biologists at SCCF and JNDDNWR have assisted in road cruising, and if they see something odd, large, or notable, SCCF receives a call or an email and the record is evaluated. Data recorded includes the date, species, and Global Positioning System (GPS) coordinates. A general rule used by many herpetologists is, "If it's present, it will show up on the road eventually." The roadkill survey is an excellent way of finding previously unknown locations of rare species. Also, after enough locality records are documented, ecological corridors—places where a high percentage of animals choose to cross roads—become evident.

Dedicated public outreach remains an invaluable way of gathering information. By giving presentations, writing newspaper articles, and just spreading the word, we have been able to gather tremendous amounts of data on the islands' herpetofauna. These activities make people more aware of their surroundings and the information derived from this discourse is more plausible—not to mention that many snakes get to live another day. For many years now, people have routinely made phone calls or sent emails asking what kind of amphibian or reptile they are looking at or is inside their house. This kind of information is important but must be filtered because people tend to make incorrect observations. This is all part of the educational process.

Sightings are classified using a documented/undocumented protocol. For a sighting to be considered documented, the animal must be photographed or brought in to be verified by someone else (usually a herpetologist). An undocumented sighting is one that is reported but cannot be proven. This process eliminates unverifiable speculation. For assembling the SCCF herpetofauna lists and this book, we used a 20-year time period to decide whether a species was extant or extirpated. If there are no documented records within the past 20 years, then we consider a species extirpated on the island. Animals can be rediscovered, however, and placed on the extant list once again.

In the few instances where herpetofaunal species were not successful at establishing themselves islandwide over time, we include an individual distribution map specifically for that species. These maps are based on collection-locality data and in some cases include species that have not been collected or otherwise validated since their original documentation. Some are assumed extirpated from the islands' herpetofaunal community. We do not include collection maps for species that are common, well represented, and generally distributed on these barrier islands, although we note the habitat types where each occurs.

It is not within the purview of this project to cover all aspects of the biology or life history of the amphibians, turtles, crocodilians, and lizards and snakes that are discussed in the following account. We write anecdotally about each in a comments section about their conservation status, ecology, population status, and the published contributions relative to their individual distribution and life history insofar as it relates to Sanibel and Captiva islands. For general information on the species included in the account, other literature sources are cited. A bibliography for further reference is appended. The World Wide Web contains much additional information on the biology of amphibians and reptiles, and it is just a mouse-click away.

~Part II~

ANNOTATED LIST OF THE AMPHIBIANS AND REPTILES OF SANIBEL AND CAPTIVA ISLANDS, FLORIDA

AN OVERVIEW

We follow established common[1] and scientific names as published in the current edition of *Scientific and Standard English Names of Amphibians and Reptiles of North America North of Mexico, with Comments Regarding Confidence in Our Understanding. Edition 6.1* (Revision dated May 24, 2011). This publication was produced as a herpetological circular by the Committee on Standard English and Scientific Names of The Society for the Study of Amphibians and Reptiles (SSAR) (Crother 2008). It is officially recognized and adopted by the Society for the Study of Amphibians and Reptiles, American Society of Ichthyologists and Herpetologists, and the Herpetologists' League. Our list has been modified to follow other editorial rules and we have opted not to use capitalization for every word in the complete common name of each species, as does SSAR.

For our readers who are unfamiliar with taxonomic jargon a brief overview of this specialized and unique system is necessary. A Swedish botanist/medical doctor, Carl Linnaeus (1707-1778), developed a method to give every biological organism, both extant and extinct, its own personal and specific scientific name. These are sometimes referred to as "Latin names," although they may be of Latin or Greek origin. In 1758 Linnaeus published the 10th edition of his monumental *Systema Naturae*, and by the time of his death this would evolve into a multi-volume work as an increasing number of naturalists around the world sent him specimens for classification, naming, and inclusion in his system. It was this 10th edition that initiated the currently used naming system. Any previously used scientific names are not valid. Common (vernacular) names of a species are not used for allocation into proper systematic classification. Local common names are often highly variable, and many snakes, for example, may be known by several different common names within their geographical range. Scientific names eliminate this confusion. Linnaeus' gift to science is known as the binomial system of nomenclature. Over time this evolved into the trinomial system. We use this approach throughout our text, although there is a trend among taxonomists to avoid trinomials as subspecies or regional variants.

The scientific name of an organism is based on a combination of Greek and Latin (usually the latter is dominant), and each name must contain two parts. The first word is capitalized and identifies the **genus**, a subgroup in a family or subfamily to which the organism is most closely related. The second or middle name is the **species**. Species is a subdivision of closely related lineages with common characteristics of a single genus as a common ancestor. This combination represents the binomial name. As systematic zoology advanced, the scientific name was sometimes modified by the addition of a third or sub-specific name to recognize morphologically distinct populations of a species. In common usage, a **subspecies** is designated by the additional name (e.g., the black racer's name would be written, *Coluber c. constrictor* or even abbreviated further as *C. c. constrictor*).

The individual species in our Account of Species section are not organized in any formal taxonomic order. Each is listed in its respective family alphabetically by genus.

In the heading for each species in our list, the first use of scientific names is followed by the name of the person who described the species and the year the description was published. In some instances the published name and year may be set in parentheses. When the describer's name and date are enclosed in brackets, this means that the original generic name has been changed over time by taxonomists and replaced as we have represented it.

We have opted to use the metric system in conveying units of measures, weights, and volumes. This is a standard in scientific use, but because this is a semi-technical work, we include the U.S. Customary Units measurements in adjoining brackets for ease of interpretation by those unfamiliar with the metric system.

Most of our photographers are professionals and have allowed us to use the finest examples of their work. In each case the name of the photographer follows the caption associated with the photograph. Wherever possible, the authors selected photographs of specimens that were taken in the field, or at the least collected on Sanibel and Captiva islands. It was not always possible, however, to photograph island specimens exclusively. Where images of off-island individuals are used, they are representative of the individual species or subspecies that is under discussion.

Although long-term sea turtle studies have been conducted on these barrier islands since 1959, a series of representative photographs of species that are known to nest or have stranded on these beaches were simply not available. We were fortunate to acquire some excellent images of both adult and hatchling sea turtles from national and international colleagues for use in this book.

In the instances where species were introduced to Sanibel or Captiva Island after record-keeping started in 1959, we divided those range extensions into two categories: **active** introduction and **passive** introduction. In this context, *active* is defined as the intentional release of nonindigenous or exotic species into island ecosystems. We use the term *passive* to define an incidental introduction that occurred without the intent of release on the part of any person (e.g., specimens traveling to Sanibel or Captiva Island along with soil such as fill dirt, in potted nursery plants, or secretively in clumped field-grown vegetation. In other passive examples, individuals arrived on the islands transported by water on their own volition by swimming across Pine Island Sound.

In our extensive Account of Species section, we have incorporated a series of range maps. These maps are useful tools and provide the reader specific locality data for the species under discussion. Our maps help document the location where the first collection or observation of a species is known to have been made on Sanibel or Captiva Island. In each case, one of us or a herpetological expert has examined voucher specimen(s), and we consider the documentation to be valid and worthy of inclusion in the book. We do not use range maps for any of the indigenous herpetofauna of Sanibel and Captiva islands that were already well established and common when this study was begun in 1959, with the exception of the pig frog (*Lithobates grylio*). A range map with this frog's known release site is provided because its release into the Sanibel Island freshwater wetland system is a matter of record. The center of each map's yellow circle indicates the collection site.

~CLASS AMPHIBIA~

Based on collection data from field sampling and long-term personal observations, we are convinced that salamanders were always incapable of making the trip across Pine Island Sound and San Carlos Bay to reach Sanibel or Captiva Island because of the salinity barrier.

Typically, the salinity of the water distributed through the bay and sound ranges between 11.5 parts per thousand (ppt [o/oo]) and 32.7 ppt (Wang and Raney 1971). This variability is caused by seasonal loads of freshwater that reach the estuary via the Caloosahatchee, as well as the Estero, Imperial, and Peace rivers and the smaller creeks that supply them.

To make the passage, salamanders would have had to cope physiologically with the high levels of salinity (Byrne and Gabaldon 2008) and would not likely be capable of reaching the shores of the barrier islands alive. Thus, none of the four[2] genera of salamanders that occur inland on the mainland behind the mangrove periphery of the Caloosahatchee estuary (*Amphiuma, Notophthalmus, Pseudobranchus,* and *Siren*) have been documented on Sanibel or Captiva Island. We do not rule out, however, that in the future humans might release some of these amphibians into the island's interior freshwater wetlands where they could survive.

> Sanibel Island is universally known as a "Sanctuary Island." Resident wildlife species, which include some reptiles, are protected by city, state, and federal regulations. These rules prohibit the collection of all turtles, crocodilians, and some snakes. Do not collect, harass, attempt to kill, or kill the herpetofauna you find living on these barrier islands. The collection of any species of these animals, both amphibians and reptiles, are strictly prohibited on public lands and those private lands controlled by conservation organizations on both Sanibel and Captiva islands. Protection is provided by city ordinances, state statutes, and federal regulations, including the Endangered Species Act. Violators will be prosecuted.

~ORDER ANURA—THE FROGS~

Acris gryllus (Harlan, 1827)
Southern cricket frog[3]

The southern cricket frog is now extirpated on Sanibel and Captiva islands. *Daniel Parker*

Other common names: Cricket frog.

Similar species: Greenhouse frog, southern chorus frog, little grass frog.

General range of species: A southeastern species, found from eastern Louisiana to southeastern Virginia. This species occupies the entire state of Florida, with the exception of some islands and keys.

Island distribution: Extirpated, known from one locality on Sanibel Island.

Preferred habitat type(s): Southern cricket frogs can be found on the mainland alongside most freshwater habitats from puddles to lakes and even rivers. They seem to prefer quiet waters. They generally position themselves in the grass or near rocks adjacent to the water's edge where they can make a quick escape to the water when threatened.

Size: A small frog, 1.5 to 3.3 centimeters (cm) (usually just less than an inch [in] long) with a maximum recorded snout to vent length (SVL) of 3.175 cm (1.25 in).

Color: Different shades of brown or sometimes gray or green. The southern cricket frog can also be a mixture of these colors, and in parts of Florida some are red.

Other characteristics: All cricket frogs have a dark triangle between the eyes on the top of their head. This triangle is also present in some greenhouse frogs. The cricket frog also shares a warty, bumpy skin with the greenhouse frog. *Acris gryllus* in Florida is distinguished from its northern counterpart (*Acris crepitans*) by possessing reduced webbing on the rear appendages, as well as longer legs. The southern cricket frog has two dark stripes on the inside of the thigh.

Diet: Insects and other small invertebrates.

Reproduction: As with *Acris crepitans*, *A. gryllus* is primarily a warm-weather (summer) breeder, but because of Florida's climate, this can happen almost every month of the year. In South Florida, choruses of cricket frogs can be heard as early as February, although early winter choruses (November and December) are not uncommon with extended rain and above-average temperatures. The female deposits eggs singularly or sometimes in small groups of three or four. These may stick together in a loose mass of seven to ten eggs per mass. Eggs often attach to submerged vegetation, but some may drop to the bottom.

This southern cricket frog tadpole was photographed in Putnam County, Florida. C. Kenneth Dodd, Jr.

Call: The call of the southern cricket frog resembles two pebbles being tapped together in a steady rhythm.

Life history: Although the cricket frog is in the hylid (treefrog) family, it is not an avid climber since it lacks the toe pads of most tree frogs. It relies heavily on camouflage and agility in its environment along the water's edge. It is a nervous frog that will quickly start jumping away to avoid danger.

Population status: Extirpated on Sanibel Island. Populations are decreasing on the mainland as a result of loss and degradation of habitat and toxins in the water at their breeding sites.

Threats: If Sanibel Island still had a population, the main threat would be saltwater intrusion into the freshwater wetlands or toxins leaching into those habitats from runoff. These frogs would probably fare quite well in the freshwater interior wetlands of Sanibel Island if individuals successfully immigrated to the island again.

Range Map 01. Sanibel Island collection site of *Acris gryllus*, *Pseudacris nigrita*, and *Pseudacris ocularis*—1959 & 1960.

Comments: The southern cricket frog was first collected on Sanibel Island in the early summer of 1959 in a seasonally flooded wetland in the island's eastern interior. The habitat where this frog was collected has since been modified because of the canalization of the Sanibel Slough and real estate development. Circa 1960, Beach Road served as an earthen dam at the eastern boundary of this freshwater system. By 1963, a primitive water-control structure was positioned at Beach Road. The design and engineering of this barrier were altered and improved through the years. It is currently a substantial structure known as the Beach Road Weir. Land to the east was also canalized when the waterfront lots with their tidal finger canals were constructed and named Sanibel Estates. Shell Harbor subdivision came later. The historical breakout outflow of the Sanibel River was eliminated, and surplus water from the island's interior poured into the tidal system via the Beach Road Weir and a larger water-control structure known as the Tarpon Bay Weir, originally installed by LCMCD in 1959. The Tarpon Bay Weir connects that bay with the interior wetlands via a canal and box culvert (Map 6).

In 1959, Las Conchas Road left Periwinkle Way and passed through the above wetlands near Beach Road. This road allowed people to access a small near-beach subdivision known as Las Conchas del Mar. LeBuff was attracted to this small wetland area one rainy night because of an exceptionally loud frog chorus (including the "advertisement" sounds produced by male cricket frogs. He managed to capture and identify several of them. Las Conchas Road has since been vacated, and the location of this marsh was replaced with the privately owned Periwinkle Park and Campground. By this time the southern cricket frog had apparently disappeared from Sanibel Island.

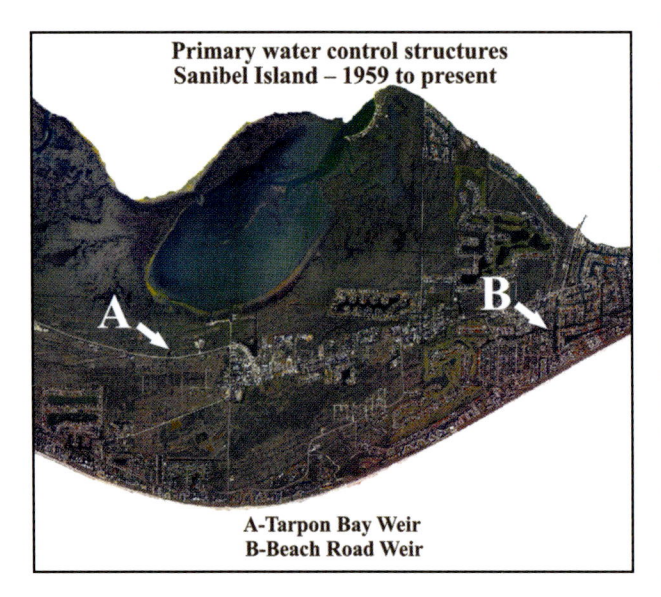

Map 6. East-central Sanibel Island in 2012. These two weirs and their predecessors were designed to control water levels in the Sanibel Slough. Since 1974 the policy of water managers has been to lower maximum levels and hold a reduced head of freshwater in the wetlands compared with what existed prior to that year. When LCMCD built the original structure, A, in 1959 they followed guidelines suggested by scientists with the Florida State Board of Health. Under that concept, water elevations were kept much higher in the interior wetlands between 1959 and 1974 than they are today. The reduction in water tables is responsible for the negative habitat changes that have occurred in the interior wetland system and the coincidental elimination or decrease in some amphibian populations after 1974.

Anaxyrus quercicus (Holbrook, 1840)
Oak toad (Passively introduced)

The tiny oak toad was collected at one location on Sanibel Island, but the species never became established. *Daniel Parker*

Other common names: None.

Similar species: Southern toad.

General range of species: A southeastern species, found from eastern Louisiana to southeastern Virginia along the coastal plain. This species occupies the entire state of Florida, with the exception of some islands and keys.

Island distribution: Extirpated, known from one locality on Sanibel Island.

Preferred habitat type(s): Primarily in upland hammocks, pine woods, and pine scrub.

Size: The SVL of this small toad ranges in length from 1.9 to 3.8 cm (0.75 to 1.5 in).

Color: Primarily brown (light to dark) on the dorsal side with a white ventral side. It has a characteristic white stripe on the middle of the back that extends to the snout.

Other characteristics: The oak toad has one pair of small elongated parotoid glands behind its eyes, as well as three to five pairs of dark spots on its back. In comparison, the parotoid glands of the larger southern toad (*Anaxyrus terrestris*) are kidney shaped and very large.

Diet: Insects and spiders.

Reproduction: The oak toad primarily breeds during the warmer months concurrently with heavy rain events. It breeds and deposits eggs in temporary wetlands such as puddles, ditches, flooded flatwoods, and marshy meadows. The female deposits short egg strings, each of which contains three to six eggs. Groups of strings may remain separate or form small clusters. The female deposits about 14 to 30 eggs per reproductive bout for a total of as many as 700 eggs per season (Dodd 2013).

This oak toad tadpole is undergoing metamorphosis and is beginning to develop its rear legs. U.S. Geological Survey (USGS)

Call: The call of the oak toad resembles that of a newly hatched chick. It is a high-pitched peep that sounds somewhat like spring peepers (*Pseudacris crucifer*) when in a loud chorus.

Life history: This is the smallest "true" toad in the U.S. It resembles the subadult stage of most North American bufonids (Family Bufonidae) and is often misidentified. This mostly diurnal toad lives in self-created burrows or under debris in uplands. Heavy summer rains trigger breeding activity by causing the oak toad to seek shallow wetlands to reproduce.

Population status: The oak toad never became established on Sanibel Island. Populations on the mainland appear to be stable in undeveloped areas.

Threats: If this toad had become established on Sanibel Island, threats would be similar to those that impact other frogs and toads. It would probably fare quite well in the uplands of Sanibel Island if individuals ever again reach the island and reproduce at a level sufficient to become established.

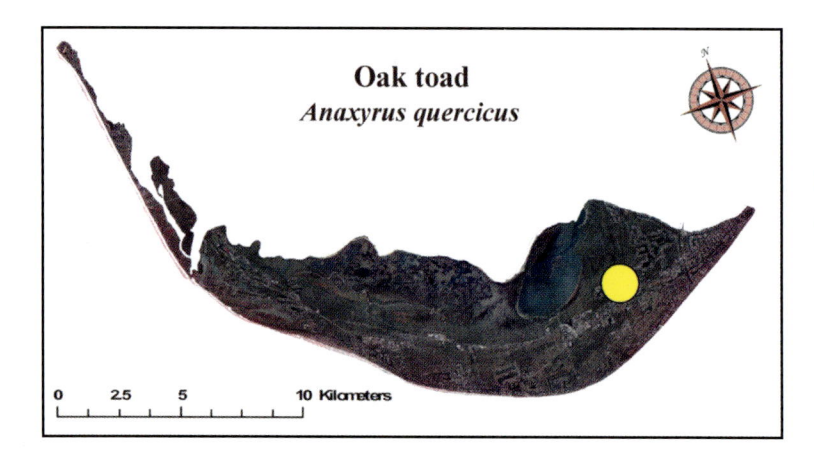

Range Map 02. Collection site of *Anaxyrus quercicus*—1982.

Comments: The diminutive oak toad did not occur on Sanibel Island in 1959. It did not make an appearance on the island until 1982 when Steve Phillips (pers. comm.) observed and collected a small number of oak toads in the Gumbo Limbo subdivision. Phillips suggested that fill dirt, which was transported from the mainland to Sanibel Island for home construction sites in this subdivision, was the carrier that brought these toads to the island. The oak toad was apparently unsuccessful in establishing itself on the island and has not been documented by either collection or voice since.

Anaxyrus terrestris (Bonnaterre, 1789)
Southern toad

An adult southern toad. The wart-covered skin and the conspicuous elongated parotoid glands behind the eye are clues to this toad's identity. *Daniel Parker*

Other common names: Common toad, toad, toad-frog.

Similar species: Oak toad.

General range of species: A southeastern species, found from eastern Louisiana to southeastern Virginia along the coastal plain. This species occupies the entire state of Florida, with the exception of some islands and keys.

Island distribution: Found island-wide except for areas with brackish and saltwater.

Preferred habitat type(s): Found throughout the island near freshwater wetlands, sandy areas, hardwood hammocks, and near human developments in both hydric and xeric habitats.

Size: Typically 6 to 9 cm (adults are primarily 2.5 to 3 in) with a maximum recorded SVL of 11.3 cm (4.44 in).

Color: Mostly gray, brown, or red-toned with or without dark blotches.

Other characteristics: The southern toad has enlarged knobs on the cranial crests between the eyes and kidney-shaped parotoid glands. The male is usually smaller than the female. The subadults can resemble an adult oak toad (*A. quercicus*) by having a mid-dorsal white stripe.

Diet: Insects, snails, spiders, and small frogs.

Reproduction: The southern toad is primarily a spring/summer breeder. Its call can be heard very early in the year if the dry season is interrupted by a substantial rain event. It lays approximately 3,000 eggs in (usually) two strings.

This larval southern toad is just beginning metamorphosis with development of its hind legs. *USGS*

Call: The call of the southern toad resembles a high-pitched constant trill. A chorus of southern toads sounds like a choir holding a single note indefinitely. Its call is often mistaken for locusts or the humming of large electric grids.

Life history: The southern toad is the southern equivalent of the American toad (*A. americanus*) from the north. These are referred to simply as "toads" by most people. It is common in residential areas and is able to sustain populations in most of these areas. It is usually nocturnal as it spends most of the day hiding under debris such as logs, rocks, or litter in self-made burrows. Contrary to long-established folklore, toads do not give human handlers warts.

Population status: The populations on Sanibel, Captiva, and the mainland appear to be stable. Many are killed on roads at night during rain events, but the huge numbers of transformed juveniles each year helps offset this loss.

Threats: The only threat to the southern toads on Sanibel Island would be reduction of freshwater wetlands due to saltwater intrusion. Sanibel Island has enough suitable habitats to sustain populations. Captiva Island has fewer freshwater wetlands than Sanibel and most properties are heavily vegetated, although there are natural and manmade depressions that hold water long enough for these toads to reproduce successfully.

Comments: The southern toad, *Anaxyrus* (formerly *Bufo*) *terrestris*, was distributed through the upland ridge system of Sanibel when LeBuff started to document the island's herpetofauna in 1959, but he never considered it to be abundant. Over time, the vegetation on the island underwent transition coincidental with the changes in the surface-water regimen because of mosquito control activities. Coupled with the rapid encroachment of pest plants, the frequency of observations was further reduced, suggesting if a very large population might have been present in the past it no longer exists.

Eleutherodactylus planirostris (Cope, 1862)
Greenhouse frog (Passively introduced)

An adult greenhouse frog. This tiny frog is not usually encountered or visible during daylight unless one lifts debris under which several dozen may be concealed. Then, they leap away so quickly they are sometimes hard to recognize as frogs. *Daniel Parker*

Other common names: None.

Similar species: Southern cricket frog, Florida chorus frog, little grass frog.

General range of species: *Eleutherodactylus planirostris* is native to the Caribbean Islands (e.g., Cuba, Grand Cayman) and the Bahamas. It has been introduced into several countries including the U.S. (including Hawaii), Mexico, Honduras, Grenada, Panama, the Caicos Islands, Jamaica, and Guam. It is suspected that the greenhouse frog arrived in these countries via potted plants from the exotic plant trade. In the U.S., it has expanded from the Keys to the Florida panhandle. It also has established populations in Alabama, Georgia, South Carolina, Mississippi, and Louisiana.

Island distribution: Found throughout Sanibel and Captiva islands in natural and manmade environments.

Preferred habitat type(s): The preferred habitat of the greenhouse frog is damp areas with plenty of cover. It is found in gopher tortoise burrows, plant nurseries, hammocks, backyards, and debris piles.

Size: Tiny, 1.25 to 3.1 cm (the adult is usually just under 1.0 in long), with a maximum recorded SVL of 3.2 cm (1.25 in).

Color: Brown or red-toned brown with red eyes and a white belly.

Other characteristics: *Eleutherodactylus planirostris* can be either striped or mottled. It has warty skin, and the dark pattern between its eyes can look triangular, resembling a southern cricket frog (*Acris gryllus*).

Diet: Insects.

Reproduction: Breeds in the warmer months (April to September).

Call: The call of the greenhouse frog is a series of quiet low chirps that resemble sounds made by insects.

Life history: The greenhouse frog is an exotic species in Florida. *Eleutherodactylus p. planirostris* uses a reproductive strategy whereby development occurs entirely within the egg; there is no tadpole stage, and the froglets hatch as very tiny copies of the adults. The greenhouse frog spends most of its time hiding under debris. It is very common in pool areas alongside private homes, hiding under pool toys, inner tubes, and furniture, but is also found in natural habitats.

Population status: The greenhouse frog's population status on Sanibel Island and the mainland is stable and more than likely spreading.

Threats: Since this is an exotic species, the threats to its existence in Florida have not been considered. The threats it poses on native fauna are minimal. With the absence of small frogs such as the little grass frogs, southern cricket frogs, southern chorus frogs, and oak toads, the greenhouse frog cannot be construed as competing for food. It simply occupies an unoccupied niche.

Comments: According to Dundee and Rossman (1989), the greenhouse frog was introduced into Florida from the Bahamas or Cuba with the assistance of the nursery plant importation trade. After establishing a Florida foothold in 1875, this tiny and unique frog spread rapidly, and today its Florida range includes most of the peninsula.

The greenhouse frog was already well established in developed areas when LeBuff first encountered several specimens underneath scrap lumber near an old farm homestead on Sanibel Island in 1959. Because of its unique life history, this frog has managed to survive on Sanibel and Captiva islands, whereas three other small and more aquatic frogs have apparently disappeared from the island.

Specimens of the greenhouse frog were collected on Sanibel Island and sent to OSU. The catalog numbers are A-622, A-678 to 684.

Gastrophryne carolinensis (Holbrook, 1836)
Eastern narrow-mouthed toad

The eastern narrow-mouthed toad congregates in summertime-flooded roadside ditches and other wetlands in breeding assemblages. The male's advertisement call resembles the noise made by a bleating sheep. *Daniel Parker*

Other common names: Sheep frog/toad.

Similar species: None.

General range of species: A southeastern species found from eastern Texas and Oklahoma to coastal Virginia and as far north as central Missouri. It occurs throughout the entire state of Florida.

Island distribution: Found island-wide in lowlands except for areas with brackish and salt water.

Preferred habitat type(s): On Sanibel and Captiva islands, the eastern narrow-mouthed toad lives near freshwater bodies, swales, and low areas in natural habitats and near human developments. It spends the day under moist logs, rocks, or debris.

Size: A small amphibian, 2 to 3.6 cm SVL (the adult is primarily 1.0 in in length). The published record SVL for this species is 3.8 cm (1.5 in).

Color: Mostly red to brown or gray with a yellow-hued dorso-lateral stripe that may not be evident in all specimens. It has a dark ventral side with white mottling.

Other characteristics: *Gastrophryne carolinensis* has a very narrow and pointed head, hence its name. It is dorsally flattened to help it hide between narrow objects. The male has a dark throat.

Diet: Insects (mostly ants and termites).

Reproduction: The eastern narrow-mouthed toad breeds primarily during the rainy season in Southwest Florida (June-September) but can start early or go late depending on rainfall. Egg counts vary considerably and average around 500. These are massed in small clumps of eggs at the water surface.

The tadpole of the eastern narrow-mouthed toad ranges between 25 and 48 mm (1.0 and 1.9 in) in total length and takes 20 to 70 days to develop from this stage into its fully transformed characteristics. Unlike most anuran tadpoles, which are typically vegetarians and eat algae, the tadpole of this species consumes zooplankton. *Dirk Stevenson*

Call: The call of the eastern narrow-mouthed toad resembles that of a lamb/sheep.

Life history: This small frog is actually not really a toad. It is part of the wide-ranging family Microhylidae, which contains the largest number of genera in any frog family. Microhylids can be either terrestrial or arboreal. Our local species, *G. carolinensis,* is terrestrial and lives close to the ground in leaf litter or under debris.

Population status: Populations on Sanibel, Captiva, and the mainland appear to be stable. This frog often ends up in backyard pools on rainy nights as it searches for shallow wetlands to breed.

Threats: The only threat to eastern narrow-mouthed toads on Sanibel Island would be the reduction of freshwater wetlands due to saltwater intrusion. There are enough suitable habitats on Sanibel and Captiva islands to sustain populations of this amphibian. It is eaten by a wide variety of snakes, birds, and mammals.

Comments: In 1959 the eastern narrow-mouthed toad was common in freshwater wetlands. Its bleating advertisement calls were part of the regular chorus of breeding anurans at those concentrated breeding sites and at the temporary pools that formed in roadside ditches during torrential rains. This small amphibian remains part of the island's herpetofauna at present. Several Sanibel specimens of *G. carolinensis* were collected on 8 September 1959 and added to the vertebrate collection of OSU. The collection numbers are A-617 to 621.

Hyla cinerea (Schneider, 1799)
Green treefrog

The green treefrog is uniquely colored to blend with its arboreal environment. This large native frog occurs throughout the barrier island wetlands. This is a very vocal amphibian during the summer breeding season when its bell-like sounds seem to bounce around its breeding sites, as numerous males simultaneously advertise their availability to mate. *Daniel Parker*

Other common names: Rain frog.

Similar species: Squirrel treefrog.

General range of species: *Hyla cinerea* is mostly a southeastern species. It ranges from east Texas to the East Coast and up the coastal plain to Delaware. In the Midwest, its range follows the Mississippi River basin to southern Illinois. It is found throughout the whole state of Florida, excluding some barrier islands and keys.

Island distribution: Found on Sanibel Island along the Sanibel River basin.

Preferred habitat type(s): The green treefrog lives in mesic (moist) habitats. It is found near wetlands such as marshes, ponds, lakes, rivers, ditches, and flooded fields.

Size: This is one of the largest native tree frogs, 3.8 to 6.4 cm (the adult is usually 1.5 to 2.25 in long), and reaches a maximum recorded SVL of 6.4 cm (2.5 in).

Color: Light to dark green on the dorsal side. It has the ability to turn brown for camouflage. The ventral side is white. It has a white or yellow stripe on both sides of its body, and some specimens can have orange spots on their back.

Other characteristics: *Hyla cinerea* has enlarged toe pads for climbing.

Diet: Insects, various invertebrates.

Reproduction: Breeds in the warmer months (May-September), with reproduction triggered by rain. In Florida the green treefrog deposits an average of 1,200 eggs in densely vegetated shallow water. Rain-filled roadside ditches are very common breeding areas for these frogs.

The tadpole of the green treefrog. These measure 40 mm (1.6 in) in total length and take 60 days to undergo metamorphosis. *Dirk Stevenson*

Call: The call of the green treefrog sounds like "*quank-quank-quank*." Some people in the South call them "rain frogs" because they usually begin chorusing right before it rains (as the humidity goes up or the barometric pressure changes).

Life history: The green treefrog is familiar in the southeastern U.S. It lives in the trees and shrubs in or near the water. It sleeps underneath attached leaves, on stems, or in tight crevices in the vegetation such as the folds of cabbage palm fronds.

Population status: The population status of the green treefrog on Sanibel Island is uncertain. There are still many areas where this frog is very common, but the presence of the Cuban treefrog on the island has definitely diminished some local populations, especially in residential areas. The green treefrog may be in slight decline due to predation and being outcompeted for refuge sites by the Cuban treefrog.

Threats: Since most of the interior freshwater wetlands on Sanibel Island are set aside for conservation, development in this frog's preferred habitat is not an issue. Its biggest threat is the Cuban treefrog, which can be much larger than the adult *H. cinerea*. Luckily, not all Cuban treefrogs reach their maximum size. *Hyla squirella* does not have the luxury of size, however, and its populations have declined accordingly.

Comments: The green treefrog was locally common on Sanibel Island when the original island amphibian and reptile list for SNWR was compiled in 1959. It is obvious that when one makes observations of treefrog concentrations on the windows of interior illuminated residences throughout the islands in the summer that the invasive Cuban species is now more abundant.

Hyla squirella Bosc, 1800
Squirrel treefrog

The squirrel treefrog is the "weatherman" of the local animal world. Typically, preceding a rainstorm, individuals of this species will call from their perches high in the canopy vegetation to announce rain is imminent. This "rain frog" is seldom wrong. It can control its pigmentation, and its body color can be quite variable, from hues of green and brown to a mottled pattern. *Daniel Parker*

Other common names: Rain frog.

Similar species: Green treefrog.

General range of species: *Hyla squirella* is a southeastern species. It ranges from east Texas to the East Coast and up the coastal plain to southeastern Virginia. It is found throughout the state of Florida, excluding some barrier islands and the Keys.

Island distribution: Found on Sanibel Island along the Sanibel River basin.

Preferred habitat type(s): The squirrel treefrog lives in mesic (moist) habitats. It is found near wetlands such as marshes, ponds, lakes, rivers, ditches, and flooded fields.

Size: Small, with a SVL of 2.5 to 4.5 cm (the adult is usually 1 to 1.75 in in length).

Color: Mostly shades of green (light to dark), although it can be gray, brown, or mottled. This frog is often hard to distinguish from the young green treefrog because of its size, and some squirrel treefrogs have a light-colored side stripe similar to that of *H. cinerea*. The white stripe of the latter, termed a "racing stripe" by some writers, is very distinctive and frequently has a dark border. *Hyla cinerea* can change its body color quickly, but the side stripe remains the same. The squirrel treefrog can also resemble the pine woods treefrog[4] (*H. femoralis*) in having a brown body color. In contrast to that species, however, the back of the thighs is a uniform gray to cream color.

Other characteristics: *Hyla squirella* has enlarged toe pads for climbing.

Diet: Insects, various small invertebrates.

Reproduction: Breeds in the warmer months (May to September) and becomes active by rain. It deposits an average of about 1,000 eggs in densely vegetated shallow water. Rain-filled roadside ditches are very common breeding areas for this frog.

The tadpole of the squirrel treefrog measures 32 mm (1.25 in) in total length. It takes 45 days for this species to transform completely to the juvenile frog stage. *Dirk Stevenson*

Call: The call of the squirrel treefrog sounds like a raspy "*quonk-quonk-quonk*." Its call also resembles the sound made by a squirrel. Some people in the South refer to them as "rain frogs," just like green treefrogs, because they usually begin chorusing right before it rains as the humidity increases and barometric pressure changes.

Life history: The squirrel treefrog can be abundant in the southeastern U.S. It lives in trees, shrubs, or in damp human structures. It sleeps underneath attached leaves, on stems, or in tight crevices in the vegetation.

Population status: The island population of *H. squirella* appears to be very small at present. Isolated animals are heard at a few locations. There are no known areas on Sanibel or Captiva islands where it is considered common. The presence of the Cuban treefrog has definitely diminished some local populations on the island, especially in residential areas. The squirrel treefrog appears to be in decline.

Threats: Since most of the interior freshwater wetlands on Sanibel Island are set aside for conservation, development in the preferred habitat of this frog is not an issue. Its only known threat is the Cuban treefrog (Campbell 1978), which is much larger than the adult squirrel treefrog.

Comments: In 1959 the squirrel treefrog was common on Sanibel Island and ranged throughout the broad freshwater wetland basin, a series of habitats that are situated between the Gulf Beach Ridge and Mangrove zones.

Osteopilus septentrionalis (Duméril and Bibron, 1841)
Cuban treefrog (Passively introduced)

An adult Cuban treefrog. Unlike most native Florida tree frogs, adults have a skin with tiny wart-like formations that can produce a poisonous secretion; the secretion is not as toxic as that secreted by the cane toad. *Daniel Parker*

Other common names: None.

Similar species: Juvenile squirrel treefrogs.

General range of species: *Osteopilus septentrionalis* is found naturally on a few Caribbean Islands (Cuba, Cayman Islands) and the Bahamas. It was first found in Key West, Florida, in 1931 and has spread northward since then. It has been reported in Georgia but has not been confirmed as breeding there. In Florida, it is found throughout the peninsula. It occurs farther north along both coasts as far as Jacksonville on the east coast and Gainesville in central Florida where it is breeding.

Island distribution: Found throughout Sanibel and Captiva islands in natural and manmade environments.

Preferred habitat type(s): The Cuban treefrog is found in all habitats on Sanibel and Captiva islands. It is an arboreal species that spends the day hiding in trees, shrubs, or crevices above the ground. It can live far from temporary or permanent water bodies but returns to them to reproduce. It can also reproduce successfully in brackish water of moderately high salinity.

Size: A "giant" tree frog, 4.2 to 12.5 cm (adults are usually 3.5 to 4.5 in long). Females are larger than males and reach a maximum SVL of 12.7 cm (5 in).

Color: Light to dark brown or gray with or without large brown mottling on the back and appendages. The venter is white-hued.

Other characteristics: *Osteopilus septentrionalis* is the largest treefrog found in the U.S., although the largest native treefrog in the U.S. is *Hyla gratiosa*, the barking treefrog[5]. Treefrogs have large expanded toe pads for climbing, and they have rough warty skin. A yellow color is usually visible on the forelimbs. They have paired vocal sacs.

Diet: Insects, various invertebrates, amphibians, and reptiles.

Reproduction: The Cuban treefrog breeds in the warmer months (May to October), and breeding is initiated by rainfall. One female is known to have deposited 16,371 eggs (Meshaka 2001) in a flattened film-like egg mass at the surface of the water.

The golden-eyed tadpole of the Cuban treefrog. *USGS*

Call: The call of the Cuban treefrog resembles the sound produced by rubbing your fingers across a balloon.

Life history: The Cuban treefrog is not an indigenous species in Florida. It is very successful at colonizing new areas, and is considered invasive.

Population status: The Cuban treefrog's population status on Sanibel Island and the mainland is stable and more than likely spreading.

Threats: The threats to the Cuban treefrog in Florida have not been considered. The threats it poses to native fauna are severe, especially to smaller amphibians and reptiles. It is known to eat other hylid (treefrog) species (e.g., green treefrogs, squirrel treefrogs) that occupy the same arboreal niche. It has been reported to replace native treefrog communities in residential areas, mostly resulting from interactions at communal breeding sites and competition for retreat sites. Even its tadpoles have been shown to be more aggressive toward native tadpoles.

Its body secretions can be irritating to some people. You will never forget the experience of handling a Cuban treefrog and then rubbing your eyes without washing your hands.

Comments: The first Cuban treefrogs on Sanibel Island were not collected until circa 1970. The species has since been very successful. Campbell (1978) recognized the potential threat to native treefrogs and commented on it. Later, Wilson and Porras (1983) substantiated the existence of this problem. More recently, Meshaka (2001) detailed this treefrog's impacts in Florida, and Lever (2003) addressed the negativity of this Cuban interloper on native hylid stocks.

Rice et al. (2011) discussed the recovery of native treefrogs in Florida after the mass removal of *Osteopilus septentrionalis* from four locations. This study carefully monitored survival, capture probability, abundance, and proportion of sites occupied by Cuban treefrogs and the native species, *Hyla cinerea* and *Hyla squirella*. Four sites in Everglades National Park were selected and capture–mark–recapture techniques were used. The sites were monitored for all species for five months, and then Cuban treefrogs were systematically captured and removed. Survival, abundance, and occupancy rates of the native treefrogs were monitored for one year after the Cuban treefrog removal started. Capture rates of native species were low prior to Cuban treefrog removal, and the estimated abundance of native treefrogs increased after commencement of Cuban treefrog removal, although abundance was found to vary with the season.

Evidence suggests that the numbers of green and squirrel treefrogs have diminished over the years on Sanibel and Captiva islands because of the invasion of Cuban treefrogs. Not only are field observations of native amphibians reduced from 1959, but the deafening choruses that native frogs produced throughout Sanibel Island's interior wetlands during the rainy seasons at that time have quieted considerably.

After a severe cold front impacted Sanibel Island in December 1988, high levels of mortality of Cuban treefrogs occurred. This was most noticeable and odiferous around structures where the frog's decomposing bodies were crowded near building soffit vents and in nearby niches where they had tried unsuccessfully to evade the lethal cold temperature.

Pseudacris nigrita (LeConte, 1825)
Southern chorus frog

The southern chorus frog is another mainland species that has disappeared from Sanibel Island. Once considered common in the wetlands of eastern Sanibel, the species was likely extirpated because of sudden habitat degradation as early as the end of 1960. *Daniel Parker*

Other common names: Chorus frog.

Similar species: Southern cricket frog, greenhouse frog, little grass frog.

General range of species: *Pseudacris nigrita* is a southeastern species, ranging from southeastern Mississippi to eastern North Carolina. It is found throughout Florida except for some barrier islands and keys. It previously consisted of two subspecies, *P. n. nigrita* (southern chorus frog) and *P. n. verrucosa* (Florida chorus frog), but no subspecies are now recognized (Moriarity and Cannetella 2004).

Island distribution: Extirpated, known from one locality.

Preferred habitat type(s): Found primarily in or near temporary water bodies such as flatwoods, ditches, freshwater swales, and puddles.

Size: A small frog, adults are 1.9 to 3.2 cm (usually around 1.0 in long) in SVL.

Color: Grayish-brown with three rows of dark marks down the back.

Other characteristics: *Pseudacris nigrita* has a black upper lip with a broken white lower lip, or a solid white upper lip.

Diet: Insects and other small invertebrates.

Reproduction: Eggs are deposited in shallow water in a loose mass. A female may produce 1,500 eggs annually, in masses of a few dozen to more than 100 eggs per mass at a time (Jensen et al. 2008).

The tadpole of the southern chorus frog averages 18 mm (0.7 in) in total length. It takes about 50 days for a larva of this species to transform into a frog. *Dirk Stevenson*

Call: The call of the southern chorus frog resembles the sound produced by moving a finger along the middle teeth of a comb.

Life history: The southern chorus frogs is typically a late winter (February and March) or early spring breeder in most of its range. In southern Florida, breeding may occur earlier depending on rainfall and temperature.

Population status: Extirpated on Sanibel Island, and populations are thought to be declining on the mainland because of habitat loss and toxic runoff. Temporary wetlands are very important for this species.

Threats: If Sanibel Island still had a viable population, this species would be stressed by the same environmental factors that impact existing populations of frogs and toads elsewhere on the barrier islands. These include surface-water runoff from asphalt pavement and fertilizers from landscaped properties. The southern chorus frog would probably fare quite well in the freshwater interior wetlands on the island if it were ever to become reestablished on the island.

Comments: Two small frogs, the southern chorus frog, *Pseudacris nigrita* (above), and the little grass frog, *P. ocularis* (below), were collected on Sanibel Island during the summer of 1960 at the same locality as the southern cricket frog, *Acris gryllus* (Range Map 01). The exact month these collections were made could not be ascertained because the file at JNDDNWR could not be located and the Cypress Lake High School preserved specimen repository no longer exists.

On September 10, 1960, a 2.7-m (8.8-ft) storm surge from Hurricane Donna crested above the Beach Road dam (before it was an engineered structure), and for several hours seawater flowed into the Sanibel wetlands. Based on vegetation dieback, it was evident that high-salinity water had reached into the system as far west as Casa Ybel Road, 1.9 km (3 mi) to the west of the breach. It is reasonable to assume that this saltwater inundation adversely impacted the amphibian population of eastern Sanibel Island and resulted in the loss of what must have been a very spatially limited population of *Acris* and *Pseudacris*. Searches for the three species since 1960 have not been successful.

Pseudacris ocularis (Bosc and Daudin, 1801)
Little grass frog

The little grass frog is aptly named. It is the smallest frog in the U.S. This is one of three species of small Florida frogs that were documented on Sanibel Island in 1959 and 1960. Soon after, they disappeared because their island distribution was apparently confined to the specific wetland system where they were recorded by LeBuff. In the latter part of 1960, a major storm overwashed their population. This resulted in their extirpation. *Daniel Parker*

Other common names: None.

Similar species: Southern cricket frog, greenhouse frog, southern chorus frog.

General range of species: *Pseudacris ocularis* is a southeastern frog that ranges from the Florida panhandle to southeastern Virginia in the coastal plain. The species occupies the entire Florida peninsula.

Island distribution: Extirpated, known from one locality.

Preferred habitat type(s): Found primarily in or near temporary water bodies such as flooded grassy meadows, flatwoods, ditches, freshwater swales, and puddles. It prefers to dwell in high grassy wet areas.

Size: Its name is descriptive of its size, and this species' SVL ranges from 1.1 to 1.7 cm (usually around 0.5 in long).

Color: Light brown to dark brown with a dark stripe through each eye that continues down the body. It can also have dark stripes down its back.

Other characteristics: The little grass frog is a long-distance jumper despite its size.

Diet: Insects and small invertebrates.

Reproduction: The little grass frog is primarily a winter breeder in South Florida but can be heard calling in the warm months, depending on rainfall. Approximately 100 to 200 eggs are deposited singly per female in shallow water and are attached to aquatic vegetation or on the submerged substrate.

Like its adult counterpart, the little grass frog tadpole is tiny, measuring 23 mm (0.9 in) in total length. Metamorphosis for this species takes 45 to 70 days. *Dirk Stevenson*

Call: The call of the little grass frog resembles the sound produced by moving a finger along the thinnest (higher-pitch) teeth of a comb. Others say it resembles a cricket.

Life history: This minute frog often goes undetected in areas because of its size and soft vocalization resembling the stridulation of crickets.

Population status: Extirpated on Sanibel Island, and populations may be in decline on the mainland because of habitat loss and toxic runoff. Temporary wetlands are very important for this species.

Threats: If Sanibel Island still had a viable population, the main threat would be development. This species would probably fare quite well in the freshwater interior wetlands on the island if it ever recolonized the island.

Comments: Refer to the southern chorus frog species account for information on the historic status of this species.

Lithobates grylio (Stejneger, 1901)
Pig frog (Actively introduced)

A pair of adult pig frogs. In most true frogs sexual dimorphism can be very pronounced and easily determined. Such is the case in this species. The tympanum, the frog's eardrum, is the flattened, external membranous disc behind and slightly below the eye. This feature clearly distinguishes the genders. The specimen to the left, with the larger ear disc is the male. *Bill Love*

Other common names: Bullfrog, southern bullfrog.

Similar species: Bullfrog, southern leopard frogs.

General range of species: *Lithobates grylio* is a southeastern frog found from eastern Texas to South Carolina along the coastal plain. It is found throughout the entire state of Florida excluding some barrier islands and keys.

Island distribution: Found throughout Sanibel Island in permanent freshwater bodies.

Preferred habitat type(s): The pig frog is found in lakes, ponds, rivers, and other permanent freshwater habitats.

Size: Large, 8.3 to 14 cm (adults are around 4 to 5.5 in long). This is the largest frog species found on Sanibel or Captiva islands, and the maximum SVL recorded for this species is 16.2 cm (6.375 in).

Color: Olive-green to brown on the dorsal side, and white with a mottled dark pattern that is more prominent posteriorly on the ventral side.

Other characteristics: A typical specimen of *L. grylio* has dark spots on its back. The male has a yellow throat, while the female has a white throat. The tip of the longest toe extends well beyond the webbing in the bullfrog, *L. catesbeianus,* and just above the webbing in *L. grylio*. The back of this frog's thighs has a row of white spots or a white line. The pig frog has one vocal sac that expands to resemble three sacs in calling males. It has a narrower snout than a bullfrog.

Diet: Insects, small invertebrates, amphibians, small reptiles, and any other live animals small enough to catch.

Reproduction: The pig frog breeds during the warmer months (April to September) in South Florida. It breeds in permanent water bodies, and the egg mass ranges from 6,000 to 34,000 eggs, with clutch size proportional to female body size (Dodd 2013).

The tadpole of the pig frog is huge, and on average measures 110 mm (4.3 in) in total length. For this species metamorphosis takes about one year. Dirk Stevenson

Call: The call of the pig frog resembles the grunt of a pig, usually in a sequence of two or three grunts. The male calls during the day or night. It is not uncommon to hear a lone male make a few grunts in the winter months on warm days.

Life history: *Lithobates grylio* is similar in many ways to *L. catesbeianus,* the American bullfrog, and to an as-yet-undescribed bullfrog that occurs in North Florida (C. K. Dodd, Jr., pers. comm.). The pig frog basically fills the niche of the bullfrog in South Florida. The pig frog's call is known as one of the iconic sounds of the southeastern U.S.

Population status: The pig frog's population status on Sanibel Island and the mainland is stable.

Threats: The pig frog does not have any serious threats to its survival on Sanibel Island. It is eaten by birds, reptiles, and fish in its environment, but the population is sustainable. On the mainland, it is hunted for food (frog legs) by people, but it is still common in most areas. "Gigging" for frog legs can put a strain on populations in localized areas.

Range Map 03. Introduction site of *Lithobates grylio*—1952.

Comments: The pig frog had apparently been unsuccessful in crossing San Carlos Bay and reaching Sanibel Island on its own before 1952. In April of that year, Refuge Manager W. D. Wood translocated 39 pig frogs from Loxahatchee National Wildlife Refuge[6] (LNWR) in Palm Beach County and released them on the BT on SNWR (LeBuff 1998). The biological reasoning for this release was to increase the diversity of wildlife prey. Over time the pig frog spread rapidly through the interior wetlands of Sanibel Island, and by 1959 this frog was being regularly harvested as food by island families. The later mosquito-control ditching program and real estate development lakes enhanced pig frog habitat substantially.

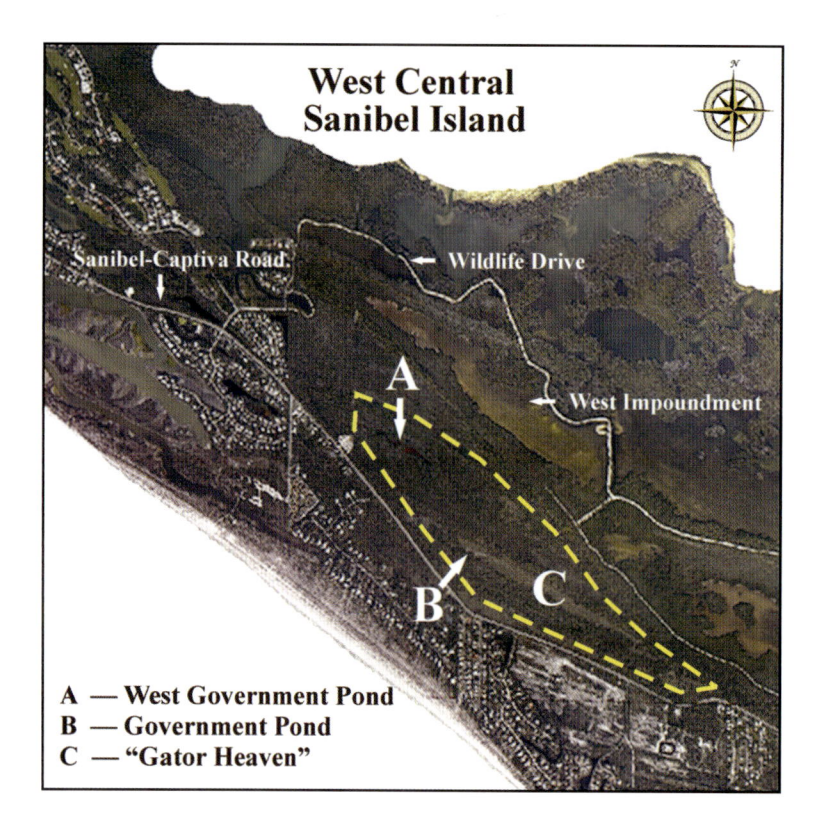

Map 7. Until the mid-60s, West Government Pond (A) was a unique habitat for the American alligator, as was the then-open *Spartina* wetland prairie called "Gator Heaven" (C) by locals who illegally harvested alligators. Government Pond (B) was also important secluded resting habitat for the roseate spoonbill when this bird was uncommon on the barrier islands.

In 1964, LeBuff discovered what appeared to be a stable colony of pig frogs living in an interior, landlocked, and impenetrable red mangrove-fringed pond, the water of which tested 40 percent of sea strength (Campbell 1978). This is known as West Government Pond and is situated on JNDDNWR (see map above). It became accessible in 1963 after a U.S. Bureau of Land Management survey line provided easy entry. This pond completely dried later that year as was its cycle and was never revisited on foot after 1967 until LeBuff made a brief and difficult walk to its fringes in 1981. The successful use of brackish water is remarkable because of the apparent inability of the pig frog to cross the salinity barrier of San Carlos Bay.

Evidence suggests this frog has the ability to adjust to variable environmental conditions. After several wet years enhanced by excessive rainfall, the flooded impounded compartment in the mangrove forest of JNDDNWR known as the West Impoundment reached a very low salinity in March 1972. By late summer of the next year, this 162-ha (400-ac) impoundment supported a large and robust population of subadult pig frogs. Subsequently their numbers wavered, but the species continued to survive. In 1979, seven permanent water-control structures were installed through the impounding mosquito dike/wildlife drive. By 1980, interchange between both impoundments and the tidal system was possible for the first time in 15 years. Thereafter, refuge staff adopted a management protocol that would permit water-quality manipulation and allow input of saline tidal waters to maintain minimal water levels over the expansive impounded flats to prevent reproduction of saltmarsh mosquitoes and provide habitat for shore and wading birds. The freshwater characteristics of this impoundment were abandoned and quickly returned to a historical estuarine system. With the return of saline water, the once-successful population of pig frogs disappeared from the West Impoundment.

Lithobates sphenocephalus (Cope, 1886)
Southern leopard frog

The southern leopard frog was once extremely abundant in the Sanibel Slough system. During summer rain events, thousands of leopard frogs leaped across the roads at night; they were ubiquitous where there was fresh water. As the system dried in the spring, and the margins of the drying swales were accessible on foot, the number of jumping leopard frogs trying to flee a person's approach was phenomenal. *Daniel Parker*

Other common names: Leopard frog, grass frog.

Similar species: Pig frog, bullfrog.

General range of species: *Lithobates sphenocephalus* is a wide-ranging frog found from East Texas to the East Coast of the U.S. It occurs as far north as central Illinois in the Midwest and up the coastal plain on the Eastern Seaboard to Long Island, New York. It is found throughout Florida with the exception of some barrier islands and keys.

Island distribution: Found throughout Sanibel Island in permanent and temporary fresh and brackish water.

Preferred habitat type(s): *Lithobates sphenocephalus* prefers to dwell near shallow water bodies. It will venture quite far from water in high grass to seek refuge.

Size: Moderate sized, 5 to 9 cm (the adult is around 2 to 3.5 in long). The maximum recorded SVL of this leopard frog is 12.7 cm (5 in).

Color: The southern leopard is brown, gray, or green with dark unpaired linear blotches that are circular or rectangular.

Other characteristics: This leopard frog has a white spot in the center of its tympanum. It has a white dorsolateral ridge on each side of the back and a long, pointed snout. The ventral side is mostly white, and there is a white line along the upper lip. The southern leopard frog has paired vocal sacs.

Diet: Insects, crayfish, and other small invertebrates.

Reproduction: In Southwest Florida the southern leopard frog breeds in late winter or early spring (February and March) when temperatures are in the upper 40s to lower 60s[7]. It calls in breeding assemblages, but it is known to call early on mornings throughout the year. It attaches egg masses containing upward of 1,500 eggs (total per female) to aquatic vegetation. The tadpole resembles that of a bullfrog but takes only three months to transform. It is known as a communal breeder because some populations of females deposit large numbers of clutches in a small area.

The tadpole of the southern leopard frog has a maximum length of 89 mm (3.5 in). It takes 90 days to complete metamorphosis. *Dirk Stevenson*

Call: The call of the southern leopard frog resembles cackling or laughing.

Life history: *Lithobates sphenocephalus* is a common frog that is often seen near the water's edge or in high grasses near water bodies. It can jump very high and far when approached and mostly jumps in different directions as opposed to straight-line jumping. In many areas, it is referred to as a grass frog because of its tendency to sit in high grasses.

Population status: The southern leopard frog's population on Sanibel Island and the mainland is stable at present.

Threats: Besides this frog's role in the food chain as prey for fish, reptiles, birds, and mammals, the Sanibel Island population of southern leopard frogs is not threatened at this time. This frog's ability to survive in brackish environments allows it to survive in water bodies where other frogs cannot, thus extending its available habitat. Off island, larger specimens are sometimes "gigged" for their meat (frog legs), and considerable numbers are collected for scientific research and teaching in some states.

Comments: The southern leopard frog was extremely plentiful in the interior wetlands of Sanibel Island in 1959. Subsistence harvesting of large adults of this abundant species for their edible legs was common among a few island families. This frog continued to be very successful in the marshes of the interior wetlands, and from all indications its extremely abundant population was stable into the mid-1960s. The southern leopard frog population has since declined from its former abundance; the days when one could walk through the shallow wetlands of Sanibel Island and have scores of these frogs simultaneously leap away in fright are gone. Lechowicz's data suggests the southern leopard frog's population may have again stabilized. He considers this frog to be relatively common in some wetland habitats on the western end of Sanibel Island.

An unquantified threat may be responsible for this frog's decline between 1965 and 2005; indeed, this threat may have affected all anurans on Sanibel Island. Earlier, we referred to the long-term chemical mosquito-control program conducted on Sanibel and Captiva islands by LCMCD. To our knowledge, there has been no definitive investigation into the impacts that the chemicals used to combat mosquitoes may have played in the reduction of amphibian populations. Elsewhere, studies have demonstrated the negative effects that mosquito adulticides and larvicides have had on a variety of organisms. Clark et al. (1987) looked into what effect thermal-fog applications of the adulticide Fenthion had on caged shrimp (*Penaeus duorarum*), and sheepshead minnows (*Cyprinodon variegatus*). The chemical was found to cause mortality among the shrimp but not the minnows. According to McAllister et al. (1999), the use of Fenthion was implicated in the decline of the northern leopard frog, *Lithobates pipiens,* and the Wyoming toad, *Anaxyrus baxteri*, in Wyoming. Fenthion has been phased out as an adulticide and is no longer applied on Sanibel Island. Naled (marketed as Dibrom), another organophosphate insecticide, is now the adulticide of choice for application where authorized on Sanibel Island by the LCMCD.

Sparling et al. (1997) studied the effects of Abate on larvae of the green frog, *Lithobates clamitans*. Tadpole mortality occurred during controlled laboratory experiments. Abate is a common anti-mosquito larvicide that continues to be part of the chemical arsenal of the LCMCD and is applied throughout Lee County, including some parts of Sanibel Island. In 2009, the USFWS prohibited the use of Abate on lands and waters of JNDDNWR. Where Abate is still applied on Sanibel and Captiva islands by LCMCD, its use is rotated with other larvicides (i.e., *Bacillus thuringiensis israelensis*, or *Bti,* a microbial insecticide, and Methoprene, an insect growth regulator), to prevent a buildup of tolerance levels or a resistance of target larvae to the active ingredients of any of the three compounds.

A much more recent study in California (Sparling and Fellers 2009) has implicated other insecticides to be chronically toxic to larval amphibians. The compounds implicated in this study were the commonly used organochlorine Endosulfan (a restricted-use pesticide) and the cholinesterase-inhibiting organophosphorous insecticides. The latter include Chlorpyrifos, Diazinon, and Malathion. Most recently, Malathion was used extensively in California's agricultural regions in the 1980s to combat the Mediterranean fruit fly, *Ceratitis capitata*. These insecticides were shown to slow development in the Pacific chorus frog[8], *Pseudacris regilla*, and the foothill yellow-legged frog, *Rana boylii*.

The active introduction of the pig frog, *Lithobates grylio,* into the interior Sanibel Island wetlands in 1954 also may have been a factor that led to the decline of the southern leopard frog in the habitat the two frogs share. According to Lamb (1984), the diet of mainland pig frogs consists of crayfish and anurans, with crayfish[9] making up 20 percent of the frog's diet.

Duellman and Schwartz (1958) list the remains of southern leopard frogs among the stomach contents of pig frogs from data they were provided by the FGFWFC. In the absence of a sufficient crayfish population, the abundant southern leopard frog may have become the pig frog's primary prey on Sanibel Island, and over time the combination of chemical mosquito control and concentrated selective predation by pig frogs could have contributed to the southern leopard frog's reduction in abundance.

Rothermel et al. (2008) suggest that the pathogen *Batrachochytrium dendrobatidis* (Bd) has been present in amphibian populations in the southeastern U.S. for three decades. In the eastern U.S. this pathogen has been found to infect larval bullfrogs at a National Fish Hatchery in Warm Springs, Georgia, but not in the similar USFWS installation in Welaka, Florida (Green and Dodd 2007). More recently, Rizkalla (2009) has found Bd to be present and has vouchered infected individual Florida cricket frogs and bullfrogs in Orange County, in Central Florida. Although this fungus is known to be present, there is no evidence to suggest it has caused a decline in amphibian populations in the Southeast.

In time, investigators will likely document the amphibian chytrid fungus among the wild populations of Southwest Florida anurans. The previous decline of the southern leopard frog on Sanibel Island suggests that the anurans of the island be included in future studies on this disease.

Rhinella marina (Linnaeus, 1758)
Cane toad (Passively introduced)

The giant cane toad has huge parotoid glands on both sides of its neck, and the left gland is clearly visible in the photograph. As a defensive behavior, these glands can secrete bufotoxin, a chemical that can have distasteful and even serious effects on an animal that picks up a cane toad in its mouth or on a child who handles a specimen and then inadvertently gets the compound in his or her eyes or mouth.
Bill Love

Other common names: Giant toad, marine toad, faux toad, neotropical toad.

Similar species: Southern toad.

General range of species: Although many think it to be an exotic, *Rhinella marina* is native to the U.S. Its natural range is Central and South America north through Mexico to South Texas.

It has been introduced and has become naturalized in parts of Central and South Florida. Isolated populations exist in Key West and nearby Stock Island. A population in Bay County in the Florida Panhandle may have been extirpated by cold weather (C. K. Dodd, Jr., pers. comm.).

Island distribution: A small breeding population was discovered in July 2013 on the eastern section of Sanibel Island.

Preferred habitat type(s): *Rhinella marina* prefers disturbed areas such as human developments, agricultural lands, and vacant lots, but it can also be found in natural areas near freshwater. It is even known to exist near brackish mangrove systems and seems to prefer habitats near permanent water bodies.

Size: Cane toads are very large, 10 to 15 cm (adults are usually 4 to 6 in long in the U.S.). The record SVL for this toad (a specimen from Suriname) is 23 cm (9.10 in).

Color: The cane toad is brown to gray with a variety of patterns. A small individual may be hard to distinguish from the southern toad because of its similar bumpy, warty skin. The warts of *Rhinella* are red-toned. Its belly is mostly white, but occasionally it may have some brown or gray coloration.

Other characteristics: *Rhinella marina* has a pair of very large pillow-like parotoid glands (see above photograph). Like some toads, *R. marina* has an elevated bony ridge known as the "cranial crest" on the top of its head. Different species have divergent formations and positions of these crests that allow herpetologists to distinguish among species that otherwise closely resemble one another. The cranial crest of the cane toad is well developed, forming ridges above the eyes. The crest ridge contacts the parotoid glands posteriorly and joins anteriorly above the toad's snout.

Diet: The cane toad's diet consists of insects, invertebrates, reptiles, and other amphibians.

Reproduction: The cane toad typically breeds in the spring and summer months (March to July), and breeding is initiated by rain. Eggs are deposited in long strings, and a large female can deposit strings containing 5,000 to 32,000 eggs in densely vegetated shallow water. Rain-filled roadside ditches are common breeding areas for these toads.

The cane toad tadpole measures 22 to 30 mm (0.87 to 1.2 in) in total length. It can undergo rapid metamorphosis, and under optimum conditions the tadpole becomes a toadlet in ten days; in some climates and habitats this can extend to as long as six months. In general, the cane toad tadpole appears nearly black, and this along with the high position of its eyes helps to identify this tadpole as to species. *David Nelson*

Call: The call of the cane toad sounds like a motor running. It has even been described as the "typical" UFO sound from 1950-1960s science-fiction movies.

Life history: The cane toad is at home in agricultural fields, empty lots, and suburban areas that are relatively close to water. It is nocturnal and has very good eyesight in order to catch moving prey. It will eat food out of pet animal dishes and will even consume the feces of other animals.

Population status: Small breeding populations of the cane toad have been observed on Sanibel Island, all near the original collection location, as of this writing.

Threats: We do not consider survival threats to invasive species; however, this species poses a threat to dogs, cats, small mammals, and fish in Florida because of its parotoid gland secretions. This usually happens when another animal puts a cane toad in its mouth. These secretions can irritate human eyes and the skin of some people.

Comments: George Campbell (1985a) reported the cane toad on Sanibel Island after specimens escaped from his live collection in the central section of the island. There is no evidence to suggest that these animals successfully reproduced and established 28 years ago. During that interval, the cane toad was never reported, observed, or heard during call surveys.

On 17 July 2013, while at an established listening station on the eastern section of Sanibel Island during a monthly frog call survey, Lechowicz, in the company of SCCF interns Stephanie Cappiolla and Tony Henahan, heard the distinctive sound of a chorus of cane toads. They had not been heard during the earlier survey in June. After completing the survey, they returned to the site with the additional help of JNDDNWR interns Cassie Cook and Sheena Wheeler. The group was successful in collecting 10 adult cane toads from a shallow (up to 17.8 cm [7.0 in] deep) temporary retention area at a condominium. A series of unidentified tadpoles were collected from the site. No cane toad egg strings were found at that time. The tadpoles were collected to be reared by Lechowicz for later identification as to species after their metamorphosis. The adult toads were euthanized and preserved for later analysis of their stomach contents.

Range Map 04. (**1**) Known escape site of *Rhinella marina* in 1985; and (**2**) collection locality of breeding adults, 2013.

After officials emailed a citywide announcement of the presence of the cane toad on Sanibel, which included a link to an audio of the toad's call, residents in the vicinity of the breeding group(s) informed Lechowicz that they had heard the toads three weeks prior to their "discovery." The origin of these toads is unknown. We are convinced the progenitors of this population were recently passively introduced through the landscape nursery industry, and they did not originate from the 1985 escapees.

~CLASS REPTILIA~

~ORDER TESTUDINATA—THE TURTLES~

Apalone ferox (Schneider, 1783)

Florida softshell

This adult Florida softshell turtle shows the species' adaptations for a fully aquatic lifestyle. A streamlined, thin-profile shell and superbly webbed feet move this turtle through the water quickly. The extended tubelike nostrils allow softshells to remain submerged but still reach the surface for air. This turtle is also extremely fast on land. It can actually run and is often difficult to catch because of its speed. When picked up incorrectly the sharp claws can inflict deep painful scratches on the hands of the captor. The Florida softshell turtle can also extend its neck and whip its head around to bite. *Daniel Parker*

Other common names: Softshell, rubberneck.

Similar species: None.

General range of species: The Florida softshell is a southeastern species that occurs mostly in Florida. It also ranges into extreme southern Alabama, southern Georgia, and southeastern coastal South Carolina. It is not found naturally in the Florida Keys and is less common in the Florida Panhandle.

Island distribution: Found throughout the island in freshwater lakes, ponds, ditches, canals, and marshes.

Preferred habitat type(s): Although it prefers freshwater, the Florida softshell will use slightly brackish water bodies. It is primarily a turtle of still waters but it does occur in moving water in certain areas.

Size: This species is by far the largest North American softshell turtle. The adult male matures at around 15 cm (5.9 in) in carapace length, with an average size of about 33 cm (13 in). The adult female matures at about 26 cm (10.2 in), with an average size of about 63.5 cm (25 in). The largest female ever documented was 73.6 cm (28.9 in) in straightline carapace length (SCL) with a weight of 43.6 kilograms (kg) (96.1 pounds [lb]). Hatchlings of this species average 38.5 mm (1.5 in) in SCL.

Color: Primarily brown with shades of gray, green, black, and even orange in some specimens on the carapace. Their plastron is white to pink. Its head is mostly greenish-brown with occasional yellow streaks or spots. Its limbs and tail are generally the same color as the carapace.

Other characteristics: The Florida softshell turtle has a very long neck and webbed feet. It has a flattened appearance, and the shell (carapace and plastron) is not hard as in most other turtles. Instead, the shell is leathery and flexible along the margins. The Florida softshell turtle has tubercles on the anterior part of the carapace. The elongated, almost snorkel-like nostrils of this genus are unique among American turtles.

Diet: The Florida softshell turtle eats snails, insects, fish, and crayfish. It occasionally eats birds, snakes, and turtles but not regularly. Plant material is eaten incidentally.

Reproduction: *Apalone ferox* deposits 9-38 eggs in a clutch with an average clutch size of 20.6. It can deposit up to seven clutches a year, but two to four is common. The diameter of the brittle-shelled, spherical eggs averages 26.0 mm (1.02 in).

Life history: The Florida softshell is fond of soft-bottomed waterways, but that does not stop it from inhabiting rocky-bottomed canals and creeks. It often buries itself in shallow water along the edges of lakes, just shallow enough to stick its long neck and nose out of the water to breathe. It is a very fast swimmer and can disappear from an area "in the blink of an eye." It has a quick and powerful bite that usually occurs when someone attempts to pick one up while it is crossing a road. For a turtle, it is extremely fast on dry land.

Population status: Perhaps the most common turtle on Sanibel Island, the Florida softshell turtle is often seen crossing roads, basking on the sides of ditches, and surfacing for air in bodies of water.

Hatchling Florida softshell turtles are colorful compared with the drab coloration of the adult. *Daniel Parker*

Threats: Humans are a major threat to this species. It is used for food in some communities. Export of wild-caught Florida softshell turtles to China has also been widespread in recent years. Unrestricted harvest led to the revision of Florida freshwater turtle regulations limiting the daily bag limit to one per day. Juvenile softshell turtles are eaten by fish, birds, mammals, and reptiles. Adults are eaten by alligators.

Comments: LeBuff did not record the Florida softshell turtle on Sanibel Island until 1965. By then, the species was regularly observed in large real estate spoil ponds. We assume *Apalone* was present at some unknown low population level in the wetlands prior to that year. The species remains generally distributed islandwide in 2012. Specimens are regularly killed when attempting to cross busy roads, and some are captured on the roads by transient service workers and killed for food.

Caretta caretta (Linnaeus, 1758)

Loggerhead sea turtle

A post-nesting *Caretta caretta* returns to the water after depositing eggs and being tagged on Wassaw Island, Chatham County, Georgia, in May 1999. *Michael G. Frick*

Other common names: Loggerhead turtle, Atlantic loggerhead.

Similar species: Only the sea turtle has paddlelike appendages completely adapted for a lifetime spent at sea, and all five species of sea turtle that occur around Sanibel and Captiva islands share this characteristic. At certain stages in the loggerhead's development, especially as a subadult, it can be confused with the Kemp's ridley sea turtle.

General range of species: The loggerhead turtle is a circumglobal species and occurs mostly in temperate and subtropical oceanic zones. It primarily occupies the continental shelves but is capable of pelagic journeys, and individuals have crossed oceans. It is not generally considered a tropical species.

Island distribution: The loggerhead occurs in the deeper bays and bayous of the estuary, throughout Pine Island Sound, and the Gulf of Mexico. The adult loggerhead has even been observed swimming in the tidal borrow canal along Wildlife Drive in JNDDNWR. When coming ashore to nest, it uses all accessible beaches along the barrier island coasts including the Bay Beach Zone "inside" Sanibel Island along San Carlos Bay, and all the way around the outer Gulf Beach Zone to Redfish Pass and beyond. Occasionally, it even crawls out and deposits eggs on the spoil islands of the Sanibel Causeway.

Preferred habitat type(s): The adult loggerhead prefers offshore areas with "live bottoms." It also frequents the passes that separate the chain of barrier islands. Rocky outcrops in the seabed provide refuges for the loggerhead, and the biologically productive bottom provides a variety of prey items. It also frequents submerged wrecks and artificial reefs offshore of the barrier islands.

Size: A very large sea turtle, the loggerhead can attain a SCL of 122 cm (48 in) and a weight exceeding 227 kg (500 lb). Larger specimens have been reported in the literature, but the validity of the records is questionable. On average, hatchlings from Sanibel and Captiva islands have an SCL of 46.65 mm (1.84 in) and weigh 19.6 grams (g) (0.69 ounces [oz]). During the early years of the Sanibel-based studies related to this species, post-nesting females were interrupted and captured when returning to the water and regularly weighed using portable field apparatus. The largest loggerhead ever measured on the island by LeBuff had a curved carapace length (CCL) of 123.2 cm (48.5 in). Later, the heaviest individual ever to be weighed on Sanibel Island tipped the scales at 161.4 kg (355 lb).

Color: An adult loggerhead's carapace is usually fouled by a covering of benthic organisms, and the true color of the carapace is difficult to determine. In the rare individuals where a thick epibiont community is not present or where the dense covering is restricted to the posterior of the carapace, the carapace is reddish brown with radiating hues of dark olive melded into this base color. The plastron and nearby soft body parts are typically cream colored. The upper surfaces of the flippers are a rich brown-mahogany, and this color is contiguous with the upper neck and top of the head. The lateral surfaces of the head are often blotched with yellow and brown-red tones. Infrequently, the adult is vividly colored, some quite more so than others. The skin of the adult specimen pictured above is exceptionally brightly pigmented.

Other characteristics: This turtle's huge head and its usually organically fouled, algae-covered carapace, often including huge barnacles[10], are typical characteristics of older adults of this species.

Diet: The loggerhead turtle is primarily carnivorous and preys on a variety of marine organisms, including mollusks, crabs, and sponges. It also consumes jellyfish, even the venomous Portuguese-man-of-war (*Physalia physalis*).

Reproduction: Copulation between the sexes occurs at sea and, as in all reptiles with the exception of a few asexual forms, fertilization is internal. The male loggerhead never leaves the water. Based on data obtained from tagged animals, most female loggerheads that nest on Sanibel and Captiva islands adhere to a two-year nesting cycle—that is, most nest every other year. Some females, however, nest on three-year cycles, and more rarely, in successive years. Each female deposits multiple nests during a nesting season and can produce between three and eight clutches of eggs separated on average by 11-night intervals. Clutch size is variable, and on Sanibel and Captiva islands ranges from 27 to 181 eggs. Evaluation of clutch-size data between 1959 and 1990 revealed an overall average clutch size of 108 eggs.

Hatchling *Caretta caretta*, from Isle of Palms, Charleston County, South Carolina, July 2010. *Barbara J. Bergwerf*

Eggs are flexible, leathery, and nearly perfectly round; the average diameter taken at diametrically opposed points are 41.1 by 42.2 mm (1.61 by 1.66 in). On average, hatchlings emerge from their nest within 55 days.

Life history: The early life history of the loggerhead turtle is not completely understood. Biologists seem to agree that after they reach the water, the scurrying hatchlings quickly propel themselves offshore during a period known as the swim frenzy. At cessation of this intensely demanding physical activity, they are believed to have reached ocean areas where they find shelter among the floating mats of the brown macroalga known as gulfweed (*Sargassum natans*). Young loggerheads do not begin to show up in local waters again until they are about half grown.

Population status: The loggerhead is the most abundant species of sea turtle. Although listed as Threatened (see below), its populations on a Florida East Coast index nesting beach (Archie Carr National Wildlife Refuge) experienced an 18 percent increase in 2011 over any year of the decade of the 1980s (Ehrhart 2011). The worldwide population centers (based on concentrated nesting data) are the western Atlantic (Florida), Oman (on the Arabian Peninsula), Queensland (Australia), southern Japan, northeastern Brazil, and the Mediterranean (Greece and Turkey). Worldwide, the loggerhead population is divided into nine distinct segments. The populations occurring in four of these delineated segments (one of which includes Florida) are considered Threatened; in the remaining five segments the species is considered Endangered.

Threats: Not too long ago, the survival of mature loggerhead turtles while they were ashore nesting on the barrier island beaches was seriously threatened by people who were intent on killing them. The loggerhead was regularly and illegally harvested on Sanibel and Captiva islands into the late 1960s. In some parts of Florida, its eggs are still sometimes harvested for food, and there is occasional evidence of an illicit trade in sea turtle eggs.

Historically, the raccoon (*Procyon lotor*) has seriously preyed on loggerhead nests each summer on the beaches of Sanibel and Captiva islands. From 1959 to 1970, raccoons opened and consumed 75 percent of the loggerhead nests on Sanibel Island (LeBuff 1990). Canine distemper reduced the island's raccoon population in 1971, and following that epidemic, an alternative food source in the form of human food waste became available for the mammals. Thereafter, fewer raccoons were observed on the beaches, and far fewer sea turtle nests were destroyed by this predator. The coyote (*Canis latrans*) has now reached Sanibel Island, and in 2011 there were many cases clearly indicating that these canines were responsible for destruction of sea turtle nests on the island beaches. In 2013 coyotes regularly depredated nests but the total number of coyote-invaded nests was incomplete at this writing.

Invertebrates, such as the native and imported fire ant (*Solenopsis* spp.) and native ghost crab (*Ocypode quadrata*) are threats to hatchling loggerhead turtles. Fire ants are responsible for the loss of neonates of all species of turtles and have serious impacts on the young of other terrestrial herpetofauna. These ants are known to decimate entire groups of preemergent hatchling sea turtles—especially those with restricted movement that have pipped the shell and are digging their way up to the surface through the sand. Once the ants have entered the egg chamber, the hatchlings have little hope of survival. Ghost crabs burrow directly into nest cavities to dismember and consume the young turtles. The crabs also intercept, kill, and eat exposed hatchlings that are scurrying across the beach on their way to the reasonable safety of the surf.

Today, very serious perils to the survival of the loggerhead turtle and its relatives exist in the marine habitat. Despite regulations that mandate that the offshore commercial shrimp fleet fit the openings of their trawl nets with turtle excluder devices (TEDs), sea turtles continue to drown in such fishing gear. Ever-increasing deep-sea, longline fishing poses a growing threat to sea turtle survival because of the turtle-attracting baited hooks. There is a possible positive aspect to this serious situation as federal rules on incidental bycatch are implemented. Circle hooks and special dehooking equipment are being introduced in the longline fishery. These, along with the use of TEDs, may reduce fishing industry-related mortalities and the severity of turtle injuries.

In addition to the threats discussed above, beachfront development has seriously impacted the loggerhead and other sea turtles. Artificial lighting along the coast (e.g., parking lots, buildings, even road traffic and beach bonfires) affects both nesting females and their extremely vulnerable hatchlings. The adults are confused by bright light when they are approaching or leaving the beach, and hatchlings become totally disoriented when their inherent sea-finding abilities are short-circuited by bright beach-facing artificial illumination.

The City of Sanibel was one of the first municipalities in Florida to create a body of rules that regulates all aspects of beachfront lighting for sea turtle protection. Today, other coastal cities and counties have lighting ordinances in place that aim to reduce the amount of illumination that reaches the beach from electric lights. Unapproved lighting fixtures present overpowering glare, an obstacle that attracts hatchlings to the wrong direction—landward instead of seaward—and can lead to deaths by dehydration or by vehicles in busy parking lots or roadways.

Comments: The North American population of the loggerhead sea turtle is currently listed as Threatened under provisions of the U.S. Endangered Species Act of 1973 (ESA). This sea turtle is the most-studied species among the herpetofauna of Sanibel and Captiva islands. Historically, this marine turtle nested on the beaches of Sanibel Island in substantially large numbers. In the late spring of 1959, the senior author started patrolling the gulf beaches of the island in an enforcement effort to protect this turtle during the brief period they were ashore nesting. He also documented annual nesting density and attempted to curtail egg/nest loss by predators. Some island residents of that era relied on subsistence take of female loggerheads from the nesting beach for their dining tables, although it was already illegal under Florida law. The collection of sea turtle eggs for subsistence was not an important part of the island's culture. Field notes from those early days are no longer complete, but the Sanibel Island gulf beach hosted several hundred loggerhead turtle nests annually. By 1967, beach conditions allowed LeBuff to extend the study area farther north an additional 4.8 km (3 mi). Even with this increase, the overall beach length (20 km, [12.4 mi]) that year contained only 125 documented nests. Nests were documented only after their actual location and counting of the clutch size.

In 1964, tagging of individual nesting female loggerheads on Sanibel and Captiva islands became part of the island sea turtle conservation program. This work was continued through the summer of 1991. At first, a small supply of Monel metal tags was furnished by the Sanibel-Captiva Audubon Society. These had originally been procured by the Society to support LeBuff's American alligator tagging activities. After these were exhausted, Archie Carr of the University of Florida furnished LeBuff with a series of similar tags in 1967.

By 1969, LeBuff's sea turtle work evolved into the not-for-profit sea turtle conservation organization, Caretta Research, Inc. The organization then opted to use a series of its own tags, electing to use a tag prefix of CR (much later, tag prefixes also included SI[11] and CL).

This long-term tagging effort on Sanibel and Captiva islands provided a database that was critical to understanding the dynamics of this small nesting population of loggerhead turtles. In 1969, LeBuff and Beatty (1971) documented multiple nesting sites for *Caretta* on the Florida gulf coast for the first time when loggerhead turtle CR-111 was encountered nesting four times on Sanibel. In 1973, another female loggerhead (tagged CR-140 in 1970) was found nesting on Sanibel Island six times. All six clutches produced by CR-140 were relocated to a hatchery compound on the beach at Point Ybel, Sanibel Island, consistent with a state-approved conservation strategy. This female produced a total of 920 normal-sized eggs that summer. This was verified when the eggs were relocated and each clutch of eggs was counted, which was a standard documentation practice (LeBuff 1990). As a point of interest for the reader, after 1973, Addison (1996) documented seven in-season nests for an individual gulf-coast loggerhead nesting on Keewaydin Island, in Collier County, about 51 km (32 mi) south of Sanibel Island. More recently, Tucker (2009) reported eight in-season nests for an individual loggerhead turtle in Sarasota County, about 80 km (50 mi) north of Sanibel. Neither Addison nor Tucker provided total egg counts for their respective multiple-nesting turtles, because not all the egg complements produced by those females in Collier and Sarasota counties were counted although they were counted for CR-140 on Sanibel Island. By 1996, the FFWCC, issuer of sea turtle permits, changed its policy and precluded sea turtle permittees from examining freshly deposited nests and counting eggs. The productivity of Sanibel's CR-140 in 1973 remains a remarkable example of the reproductive capability of the loggerhead turtle.

Tagging also provided an overview of this species' marine distribution, as individuals moved to foraging habitat after nesting on Sanibel and Captiva islands (LeBuff 1990). Tagging data has suggested that the loggerhead occasionally does not return to previously used nesting beaches, but instead moves considerable distances away to nest. For example, female loggerheads nesting on Sanibel have been observed nesting on the East Coast of Florida (LeBuff 1974).

As with other sea turtles, the number of nesting females in any population varies annually. These are cyclic creatures that are regulated in their life activities by a variety of factors, many of which are not fully understood; therefore, ascertaining the total number comprising a specific population is difficult. In 2013, SCCF sea turtle biologist Amanda Bryant (pers. comm.) reported that a total of 348 loggerhead nests were deposited on Sanibel Island within the area LeBuff included in his 1959-1991 study beach—an increase over the summer of 1970 when the lowest number of total nests (70) was tallied for this species on Sanibel Island. An additional 130 loggerhead nests were documented on Captiva Island in 2013.

Chelonia mydas (Linnaeus, 1758)

Green sea turtle

A subadult *Chelonia mydas* browsing on turtle grass (*Thalassia*) in 1.5 m (5 ft) of water in the Hol Chan Marine Reserve, Belize, on June 29, 2011. *Sarah Gulick*

Other common names: Atlantic green turtle.

Similar species: The green turtle is similar in appearance to other sea turtles, except the leatherback.

General range of species: *Chelonia mydas* is a cosmopolitan marine species that inhabits the global tropical and subtropical seas. Nesting generally occurs between 30° south latitude and 30° north latitude. Concentrated centers of the green turtle's nesting range include Central America, the Caribbean, around the Caribbean Sea, Oman on the Arabian Peninsula, southern Japan, the Turtle Islands in the Philippines, Hawaii, and Australia. Many nesting populations of this species have been extirpated (Cayman Islands, the Florida Keys [including Dry Tortugas], Bermuda), and others are severely threatened.

Island distribution: Seasonally, subadult green turtles frequent the shallow marine systems of the CHES usually beginning in the early spring and continuing during the summer months. After a prolonged period of time during which no documented green turtle nestings occurred, females are again nesting on the beaches of Sanibel and Captiva islands.

Preferred habitat type(s): The green turtle prefers shoal inshore and estuarine areas that support good foraging habitat. For nesting, it prefers high-energy, high-dune profile beaches.

Size: The green turtle is the second largest of the sea turtles, capable of reaching an SCL of 153 cm (60 in). Hatchling green turtles have an average SCL of 50.0 mm (1.97 in) and weigh 25 g (0.9 oz).

Color: Hatchlings are dark gray-green to black dorsally and white ventrally. Subadults are attractively colored with shades of white, yellow, green, and gray. In all stages of growth, the streamlined shell is usually uncluttered with the epibionts that so commonly pervade the carapace of the loggerhead.

Other characteristics: The nesting crawls or tracks left by the green turtle can be distinguished from those of the loggerhead because the green turtle uses a different method of locomotion. A loggerhead crawl has alternating flipper marks, whereas that left by *Chelonia* leaves synchronous flipper imprints. The green turtle also excavates deeper pre-nesting body pits, and the female positions her rear flippers together over the egg chamber as if to conceal egg deposition. The loggerhead does not display this behavior and keeps its rear flippers well apart as its complement of eggs falls into the flask-like cavity it has dug in the sand.

Diet: Hatchling and small subadult green turtles are omnivorous and consume a variety of waterborne invertebrate and vegetative prey. Upon reaching maturity, the dietary intake changes, becoming almost exclusively herbivorous. In prime foraging habitat it grazes the extensive undersea meadows and consumes eelgrass (*Zostera marina*), manatee grass (*Syringodium filiforme*), and the aptly named turtle grass (*Thallasia testudinum*).

Reproduction: Fertilization occurs in the water and only the female ventures ashore. In remote Pacific Ocean populations, both sexes actually do leave the water to bask on the beaches of French Frigate Shoals, part of the Hawaiian chain of oceanic islands, and on beaches in the Galapagos. The female green turtle deposits multiple clutches, and up to nine nests have been recorded for individuals in the Western Hemisphere. Green turtle eggs are slightly larger than loggerhead eggs, are pliable and round, and have an average diameter of 46.0 mm (1.8 in). The average clutch size for green turtles nesting on Sanibel and Captiva islands is 104.7, with a range of 49 to 139 (A. Bryant, pers. comm.).

Life history: Information on the life history of the post-hatchling green turtle is incomplete. Like the loggerhead, it disappears after hatching and does not begin to show up in the local population until it reaches about 30.5 cm (12 in) in carapace length. Some areas on Florida's East Coast, such as the Indian River and Mosquito Lagoon, are recognized as important developmental habitats for this and other sea turtles. There is a significant amount of information that pertains to the incidental catch of immature green turtles in the waters of the CHES. Cooperative fishers in the former gill-net fishery once regularly reported catch of green and other sea turtles to LeBuff between 1960 and the mid-80s. Based on those records, we suspect that the CHES functions in much the same role as a sea turtle nursery, and that its waters are important developmental habitats for immature sea turtles.

A hatchling *Chelonia mydas* as it moves across the Sanibel beach, July 26, 2011. *Chris Lechowicz*

Population status: *Chelonia mydas* is making a rather dramatic comeback in Florida. That the green turtle is nesting on the beaches of Sanibel and Captiva islands tells us the population of mature green turtles is larger than it was when our study was launched. When we review the conditions of today, such as availability of nesting habitat and other environmental parameters, and compare them with the conditions of 55 years ago, we are unable to suggest any other reason for increased nesting by the species. Perhaps the good work of sea turtle conservationists is paying off, and the result is a population increase in Florida waters.

Threats: The herpes virus that results in fibropapilloma tumors is having serious impacts on green turtles worldwide. These tumors are eventually fatal because of secondary effects, although the tumor itself is considered benign. Tumor growth usually debilitates infected turtles over time, even when veterinarians have removed the tumors surgically. Severely fibropapilloma-diseased green turtles have been documented in the Sanibel-Captiva population.

Beachfront development and coincidental nighttime public use continue to be threats to nesting sea turtles on Sanibel and Captiva. The popularity of nighttime outdoor activities, such as seashell collecting or surf fishing, increases with the tourist influx at resorts and hotels. The use of bright stationary lanterns by fishermen and the hand-held lights used by shellers at the water's edge continues to impact sea turtles negatively. Many turtles intent on nesting, even those already up on the beach and well into the process, can be frightened away or abort egg deposition because of the activity of people, most of whom know nothing of the serious deleterious impacts their recreational activities create for these imperiled animals.

Comments: The green turtle occurs in the marine waters adjacent to Sanibel and Captiva islands on a regular basis, and all size-classes are represented. The Florida population of the green sea turtle has been listed as Endangered since 1978. A living green turtle was first documented on Sanibel Island at the SILS beach on Point Ybel in October 1959 (LeBuff 1990). Since then, stranded and diseased[12] green turtles have been collected on Sanibel Island and on the nearby spoil islands of the Sanibel Causeway. Ralph Woodring (pers. comm.) continues occasionally to capture small specimens of the green turtle during his bait shrimp operations in Tarpon Bay or the contiguous waters of Pine Island Sound.

In 1998, the unexpected happened when a green turtle left the Gulf of Mexico and deposited an egg complement on western Sanibel Island. A newspaper account of the event appeared in the weekly edition of the *Sanibel-Captiva Islander* on 25 September 1998. Hatchlings successfully emerged on 19 September, and the nest was opened for examination and documentation of its emergence on 22 September by SCCF volunteers. The cavity contained at least 43 empty eggshells—evidence that a contingent of the hatchlings had managed to leave the nest. In addition, there were 67 unhatched eggs, five dead hatchlings incompletely out of their eggs, and one live hatchling still buried in the nest.

Range Map 05. Location of the first successful nesting by *Chelonia mydas* ever to be documented on Sanibel Island—1998.

LeBuff (1998) discussed the relocation of two clutches of green turtle eggs from the Atlantic beach at the John's Island Club, near the coastal community of Indian River Shores, Indian River County, Florida, to the Sanibel Island beach. These jeopardized eggs were reburied within five hours of their deposition. The two clutches of green turtle eggs, containing 134 and 146 eggs, were moved from the eroding toe of a group of experimental sand-filled beach armoring devices and flown to Sanibel Island by Edward J. Phillips on 23 July 1976. LeBuff reburied them on a preselected section of well-drained, high-profile, Sanibel Gulf beach.

On 20 September, beach surface disturbance at the site indicated both complements had produced hatchlings. Nests were excavated two nights later and were found to have produced a total of 256 hatchlings. Tracks remaining on the beach face between the nest and the water indicated they had successfully made the trip across the beach and reached the Gulf of Mexico. There was no evidence that predatory raccoons had intercepted them.

Twenty-two years later, a green turtle nested for the first time in many decades on Sanibel, and the species has nested on the island most years since. In 2013, there were 21 green turtle nests on the Sanibel beach and two on Captiva. In 2012, only one lone green turtle nested on Sanibel and to the east of Knapp's Point. This is the first time this species is known to have nested on the eastern section of the island's beach (A. Bryant, pers. comm.). All other green turtle nests have been deposited on the higher-profiled beach west of Knapp's Point. Edward J. Phillips (pers. comm.) believes his green turtle egg relocation effort is responsible for the sudden appearance of this species on Sanibel Island. We will address this issue later in the section pertaining to the leatherback turtle.

It is noteworthy to mention that in 2013 Florida experienced a remarkable increase in the number of green sea turtle nests. Biologists estimate that green turtle nesting more than doubled in the state in 2013—most dramatically on the beaches of Archie Carr National Wildlife Refuge.

To address the origin of Sanibel's green turtles, it would be useful to collect DNA samples from the small group of green turtles that now nest on Sanibel Island. Results should be compared with the DNA of green turtles that nest on the beach at John's Island Club, Indian River County, Florida. The results might indicate that the two groups are genetically related.

Chelydra serpentina (Linnaeus, 1758)

Snapping turtle

A medium-size adult snapping turtle. Over time the carapace becomes smooth as the rough texture diminishes with growth. Because of the snapping turtle's sometimes sedentary lifestyle, its carapace may be covered with a lush growth of algae. This species is aptly named for its often aggressive defensive behavior; it is capable of extending its head and biting painfully in an instant. *Bill Love*

Other common names: Florida snapping turtle, snapper, alligator turtle.

Similar species: Alligator snapping turtle (not present on Sanibel-Captiva).

General range of species: *Chelydra serpentina* is an eastern species of aquatic turtle that was naturally prevented from occupying the western U.S. because of the barrier presented by the Rocky Mountains. Small populations have been established in California, however, as a result of released pets. This turtle is found as far north as southern Canada (Nova Scotia) and ranges south to the Rio Grande, in Texas. It occurs all along the Eastern Seaboard to southern Florida. The Florida snapper was formerly considered to be a subspecies known as *C. s. osceola*. This morphotype is found from the Okefenokee Swamp in southeast Georgia to the southern tip of peninsula Florida.

Island distribution: This turtle is found only on Sanibel Island in freshwater habitats.

Preferred habitat type(s): *Chelydra serpentina* is found in almost any freshwater body with a soft bottom. This can range from lentic (still) to lotic (moving) water systems. It is able to live in water with very little oxygen because of eutrophication (a process resulting from an overabundance of plants, usually due to added nitrogen from fertilizers). It prefers shallow water and will dig deep into submerged mud.

Size: The snapping turtle is large, with adults usually ranging from 18 cm to a record of 43.8 cm (7 to 17.25 in). The average weight of this species is 5.5 kg (12.1 lb), with a maximum of 21.6 kg (47.5 lb). The average SCL of hatchling *C. serpentina* is 31.0 mm (1.22 in).

Color: Primarily black, brown, or gray (carapace, head, limbs, and tail). Its plastron is black.

Other characteristics: The snapping turtle has a very small, hingeless plastron that leaves it vulnerable to predators while nesting. The carapace of the younger turtle has three low keels (ridges), which usually become less prominent and even completely disappear with age. One of the features used to classify the former subspecies, *C. s. osceola*, was the presence of pointed tubercles (projections) on the neck and a pair of chin barbells. Taxonomists continue to study geographic variation in this species in an attempt to understand its proper classification.

Diet: The snapping turtle is omnivorous. It consumes primarily fish, amphibians, invertebrates, and aquatic plants.

Reproduction: This turtle deposits 2-28 eggs in a single clutch. Its eggs are pliable, usually perfectly spherical, and average 28.0 mm (1.1 in) in diameter. Specimens found as far south as Lee County, Florida, can produce more than one clutch of eggs per year, but one clutch is common for females throughout most of this species' range.

Divergently colored common snapping turtle hatchlings. On the left is a normal-colored neonate and to the right is an albino specimen. *Bill Love*

Life history: The snapping turtle is a misunderstood species. The name and appearance give it a menacing reputation. It is quite docile in the water but earns its common name by its attitude when provoked on land. This species spends much time walking on the bottom of shallow water bodies. It digs under submerged debris to camouflage itself so it can ambush prey (a sit-and-wait predator). It waits patiently for fish or other prey to move past and then shoots its head out with lightning-like speed. This species can live more than 50 years.

Population status: The snapping turtle is not currently common on Sanibel Island. It is occasionally seen walking across roads, but its population may be greater than its visibility suggests. More sampling along the Sanibel River corridor may provide answers to questions relative to the population dynamics of this and other freshwater turtle species. The snapping turtle is seen mostly on the islands' golf courses near the water features.

Threats: Humans are a big threat to this species, and the snapper has been under pressure for its meat for quite a long time everywhere it occurs. Recent legislation in Florida prohibits the collection of this species for any purpose because of its similarity to the protected alligator snapping turtle (*Macrochelys temminckii*), which does not occur on Sanibel-Captiva. Juveniles are eaten by fish, birds, and mammals.

Comments: *Chelydra serpentina* was first documented on Sanibel Island in 1960 when six hatchlings were discovered in a tight group on the completely flooded roadbed of Island Inn Road. The site, near the northwest corner of the 100-acre BT of JNDDNWR, was covered with 25 cm (9.8 in) of freshwater. Since then, many specimens, some large, have been encountered crossing island roads. Recreational hook-and-line fishermen regularly caught them in the BT. Between 1963 and 1990, hundreds of large snapping turtles were caught in the wetlands by fishermen or picked up from the roads by motorists as road traffic and island visitation soared. Some were humanely released or rescued and taken off the road, but most were killed for human food.

Deirochelys reticularia chrysea Schwartz, 1956

Florida chicken turtle

The adult chicken turtle shares similar color characteristics with the cooter and slider, but the pattern of its carapace is different, as is the general shape of its shell and head. *Bill Love*

Other common names: None.

Similar species: Peninsula cooter, Florida red-bellied cooter, yellow-bellied slider.

General range of species: *Deirochelys reticularia* is mostly a southeastern species but does occur as far west as eastern Texas, Oklahoma, and southern Missouri. The chicken turtle occurs as far north as southeastern Virginia and south to the tip of Florida, but it is not found in the Florida Keys.

Island distribution: The Florida chicken turtle once lived throughout the island in temporary freshwater wetlands. Currently, this turtle's distribution on Sanibel Island seems to be sporadic, and it no longer frequents habitat where it once was regularly collected or observed.

Preferred habitat type(s): *Deirochelys reticularia* prefers the shallow water of swales, ditches, canals, and other temporary wetlands.

Size: The male chicken turtle averages about 19 cm (6.5 in) in SCL, and the female averages about 22.8 cm (9 in) in carapace length. The maximum recorded SCL for this species is 25.4 cm (10 in). The average SCL of hatchling *D. r. chrysea* is 29.0 mm (1.14 in).

Color: Black, brown, or green carapace with a yellowish web-like overlay pattern. The plastron is yellow with black markings on the bridge (the connection between the carapace and the plastron). The head has yellow streaks, while the limbs have one wide yellow line.

Other characteristics: The chicken turtle has a very long neck. When basking near other hard-shelled freshwater turtles, the chicken turtle is very obvious to identify because of its pear-shaped shell, as well as extremely long neck. It has striped pants (vertical stripes) on the posterior of its thighs.

Diet: The chicken turtle eats crayfish, amphibians (larvae and adults), and insects.

Reproduction: *Deirochelys reticularia* can deposit 2-16 eggs in a clutch, and it deposits one to four clutches of eggs in a season in South Florida. It is not uncommon for the chicken turtle to deposit eggs in the fall (October and November), as well as in the spring. Its eggs have been reported to take as long as 18 months to hatch in the Carolinas, where they are known to go through an embryonic diapause. The eggs of the Florida chicken turtle have pliable shells; they are oblong and average 20.0 (width) by 34.0 mm (length) (0.78 by 1.3 in).

It is often difficult to identify some hatchling freshwater turtles. Each has its own distinguishing features that aid in identification. The hatchling chicken turtle pictured above has the reticulated carapace pattern that is retained into adulthood. When viewed from behind, this species has vertical stripes adorning the rear legs. *Bill Love*

Life history: The Florida chicken turtle has a much different life history than the other turtles that resemble it, such as the cooter and slider. The chicken turtle regularly migrates from wetland to wetland. During dry conditions, it aestivates on land in self-dug burrows or in the mucky bottom of temporary wetlands. The chicken turtle is a relatively short-lived turtle, rarely living more than 10 to 15 years. The cooter and slider can live more than 30 years.

Population status: *Deirochelys reticularia* may be close to extirpation on Sanibel Island. Prior to 2012, the last record of this species is from March 2009 when Amanda Bryant and Chris Lechowicz found a recently dead specimen on Frannie's Preserve (SCCF) in a temporary wetland. Before that, the last record was from the late 1980s. LeBuff collected a young male chicken turtle in Sanibel's midsection on 23 September 2012. The chicken turtle is present in small numbers and requires further study.

Threats: The chicken turtle is preyed on by birds and mammals. Roads are a major cause of mortality because this turtle has a tendency to dwell in roadside ditches, then eventually moving from ditch to ditch, and crossing roads in the process. Roadkills are believed to be the main cause of the disappearance of this once-common turtle on Sanibel Island. It is also collected for the pet trade since it demands somewhat high prices. It is eaten by people in certain parts of their range.

Comments: Once a relatively common species throughout the interior wetland basin of the Sanibel Slough (in particular roadside ditches and older shallow real estate borrow ponds), the Florida chicken turtle is now considered rare on Sanibel Island. Through time the chicken turtle also may have fallen victim to the loss of prime habitat on the island. The temporary water bodies that were regularly in place in the interior wetlands system were altered by LCMCD in the mid-1960s. The mosquito-control operations were beneficial overall for wetland habitat enhancement, but one impact was the change to the wetland basin's historical hydroperiod. System wide, the swales that temporarily flooded to provide this turtle's preferred aquatic habitat have since been united with the permanent but still seasonally variable levels of the Sanibel River.

On 23 September 2012, during a routine weekly herpetofaunal survey on the eastern end of Island Inn Road, Sanibel Island, LeBuff encountered a young male chicken turtle in the act of crossing the unpaved road to the north of the BT. The individual was collected alive and turned over to Lechowicz for further study and release. The turtle measured 9.4 cm (3.7 in) SCL.

Dermochelys coriacea (Vandelli, 1761)
Leatherback sea turtle

This female leatherback sea turtle has just completed nesting on the beach at Sebastian Inlet State Recreation Area (southern section), in Indian River County, Florida. This specimen was tagged by the Caretta Research, Inc., tagging team led by Edward J. Phillips on 15 July 1975. She then bore tag number CR-1976 (the tag is visible on her front left flipper), and it is from her clutch of eggs that the four hatchlings of note (see page 99) were produced. This leatherback had a curved carapace length of 160 cm (63 in). *Carl Warsinski*

Other common names: Atlantic leatherback turtle, leatherback, trunkback turtle.

Similar species: Other sea turtles have morphological similarities, but with their well-armored boney shells they all are quite unlike the leatherback. Those that are hard-shelled are placed in their own group, the Family Cheloniidae. The leathery, ridged shell of *Dermochelys* is unique among turtles and distinguishes it from the other sea turtles. The leatherback alone has been assigned to the Family Dermochelyidae.

General range of species: The leatherback is also a circumglobal species and is more pelagic in its habits than the other sea turtles. It ranges through the deep blue waters of the planet's tropical and temperate seas, and is even known to range regularly into the chilly waters near Newfoundland and Argentina during its travels about the world's oceans. Its primary nesting beaches are located along the northern South American and the West African coasts. Many leatherback populations, such as those in Malaysia and along the west coast of Mexico, have declined drastically in recent decades. Nesting on Florida's Atlantic coast has continued to increase in the past few decades.

Island distribution: Generally, the leatherback stays offshore in deep water (>11 m [6 fathoms]) where it is infrequently observed by boaters at sea, sometimes within 8-16 km (5-10 mi) of Sanibel and Captiva.

Preferred habitat type(s): The leatherback is a pelagic, deepwater species and its habitat is the total oceanic water column down to considerable depths. It occurs from the surface to a recorded depth of 1,000 m (3,280 ft). Its preferred nesting sites are composed of soft sand. Even small stones or seashells will cause injuries to the female leatherback as she hauls her heavy body across a beach. It is likely that any leatherback that comes ashore on the often rough seashell-cluttered beaches of Sanibel or Captiva will receive flipper and plastron lacerations.

Size: *Dermochelys* is the largest living turtle and arguably the heaviest creature[13] among the extant reptiles. The largest known animal[14] had a curved carapace length of 256.5 cm (101 in) and a weight of 914 kg (2,016 lb). At the time of emergence from their nest, hatchling leatherback turtles from a Caribbean study beach averaged 62.8 mm (2.47 in) in SCL and weighed 40 g (1.41 oz).

Color: The adult leatherback is generally dark in color with a carapace and upper body parts that are brown-black in varying hues. An overlay of randomly positioned patterns of yellow to white blotching also appears on the upper shell and dorsal side of the flippers. The plastron is usually white-toned, and this turtle has a small, strange pink blotch atop its head. The presence, position, and benefits of this unique marking are not as yet understood.

Other characteristics: Unlike the other sea turtles, the flippers of the leatherback turtle lack claws. The unique shell of this turtle also sets it apart from all other living turtles. Rather than the normal turtle's characteristic shell, it consists of small mosaic-like bones that are held together by the leather-like epidermis encasing them.

Diet: Primarily, and naturally, jellyfish. Specimens are known to be attracted to and take baits of the longline fishing industry.

Reproduction: Copulation occurs at sea, and in Florida leatherback nesting begins earlier in the year than does that of most of the hard-shelled sea turtles. On Florida's beaches the leatherback nesting season ranges from March to July, but nesting is known as early as February. Like other sea turtles, females of this species deposit multiple nests on 10 night intervals, and each female deposits five to seven clutches of eggs during any one of her nesting seasons, which usually occur two to three years apart. Multiple nesting, a behavior common in a number of turtles, refers to a series of nestings made by individual turtles during the same or current nesting season. Each female returns to the beach periodically to re-nest on an individually regulated internesting interval. The overall span of time that is required for egg maturation, until a repeat clutch deposition occurs (the internesting interval) varies according to species.

Similar in character to other sea turtles, leatherback eggs are soft with a leathery shell but are at least as large as a billiard cue ball in size; some approach the size of tennis balls in diameter. A female leatherback produces a complement of 50 to 170 eggs with an average clutch size of 85 (excluding the undersized eggs most clutches contain). Normal-sized eggs in a leatherback egg clutch average 55.0 mm (2.165 in), the largest eggs among the sea turtles[15].

This hatchling *Dermochelys coriacea* has just left its deep egg cavity and was photographed on 8 August 2011 on St. Croix, U.S. Virgin Islands. *Aaron Garstin*

Life history: During embryological development of all sea turtles, an individual hatchling's sex determination occurs during the second trimester of an egg's incubation. Studies have shown that the pivotal temperature that determines sex in an egg complement, and produces 50 percent of each sex, is generally 29°C (82°F); however, there is variability in the transitional range of temperatures between species and their respective populations. The consensus among sea turtle specialists is that all sea turtle hatchlings that have incubated at 28.75°C (83.75°F) *or less* will be males, and all eggs that have incubated *above* 29.75°C (85.5°F) will result in females. As is true for all sea turtles, the early life history of the leatherback is still incompletely understood.

Population status: No good data exists to indicate how many leatherbacks may be near Sanibel and Captiva at any point in time. Worldwide, *Dermochelys* is considered to be endangered, with an estimated global population of fewer than 35,000 nesting females.

Threats: Marine debris, chiefly discarded plastic products, is a major threat to all sea turtles. All are known to ingest plastic, although, the leatherback may be more seriously impacted by these articles than the others. Floating plastic bags are commonly misidentified by this turtle as its preferred food, jellyfish. Offshore longline fishing is also a major threat to these unique sea turtles.

Comments: The leatherback sea turtle is Endangered as provided by the ESA and was listed as such in 1970. Until 2009, the leatherback was known on Sanibel and Captiva islands from one stranded specimen found in 1943, and a lone non-nesting emergence on the western Sanibel beach in 1988 (LeBuff 1990). On 3 June 2009, a female leatherback landed on a resort housing-developed but reasonably dark beach immediately to the west of Sanibel Island's Knapp's Point; there she deposited 106 normal and 19 yolkless eggs. The latter are typically undersized and consistent with normal leatherback egg complements (Pritchard 1971). This landing was not observed by anyone, and the nest was documented after dawn by the sea turtle staff and volunteers of SCCF, who have monitored sea turtle nesting on Sanibel since 1992. The configuration of the crawl and nest was difficult to recognize, and all involved were unable to positively identify the nest as to species. Photographs of the turtle's nest site and crawl were submitted to the sea turtle specialists of the FFWCC for positive identification. They mistakenly identified it as that of a green turtle.

Range Map 06. Location of the first known successful nesting by *Dermochelys coriacea* ever documented on Sanibel Island—2009.

A correct identification was made on 3 August 2009 when the nest was excavated for inspection by the SCCF team. This inspection is part of standard post-emergence protocol for every sea turtle nest to evaluate hatching success. Examination of the nest contents is carried out at a predetermined length of time after most hatchlings have vacated the egg cavity. In this case, hatchlings were still buried that had not exited with the main sibling group. These were photographed for documentation prior to being released. Other than the empty eggshells and infertile eggs, four leatherback hatchlings were discovered alive in the nest along with an egg containing a partially developed embryo (A. Bryant, pers. comm.).

In 1975, on a whim and without specific authorization, Edward J. Phillips, special projects director of Caretta Research, Inc., removed four viable eggs from a leatherback nest on the Atlantic Ocean beach of Sebastian Inlet State Recreation Area in Indian River County, Florida, while he watched the deposition of the eggs. At this time, Phillips was legally conducting sea turtle studies on the park beach. All four eggs successfully hatched and were reared in natural seawater in his middle-school classroom aquaria (Phillips 1977).

In June 1976, Phillips returned for his customary summer work with sea turtles. Three of the leatherbacks had survived and were brought to Sanibel Island where they were released into the Gulf of Mexico at Point Ybel[16]. Their sexes could not be ascertained at the time, and as they moved on their own across the dry beach they randomly hesitated frequently, but upon reaching the water and experiencing their first immersion in water that was unobstructed by glass, they swam off and disappeared.

According to Phillips (1977), the four eggs from which these turtles hatched were incubated at 29°C (85.2°F). More specifically (E. J. Phillips, pers. comm.), the eggs developed in a beach-sand-filled container inside an incubator that was thermostatically controlled. This incubation temperature was close to the upper range of the pivotal temperature at which females develop. Is it possible that one or all three of these leatherback yearlings were females, and did they imprint while crossing the Sanibel Island beach or at their point of entry into the surf? Did one female survive and return to the island to nest successfully 33 years later? Both are interesting questions. Although we recognize that some sea turtle biologists will first consider theoretical clutch-cohort survival models, we offer the same suggestion relative to the green sea turtle egg relocation success story discussed earlier, on page 87.

Little is really understood about a neonate sea turtle's possible imprintation to a natal beach, although molecular data support the concept that tight nesting assemblages are consistent with natal homing (C. Kenneth Dodd, Jr., pers. comm.). Although there is not yet any conclusive evidence to fully support the belief that following maturation female sea turtles return to their natal beach, there is almost universal acceptance of such a remarkable feat by biologists.

Eretmochelys imbricata (Linnaeus, 1766)
Hawksbill sea turtle

A handsome subadult *Eretmochelys imbricata* cruises a rocky-bottom section of a reef in the Atlantic Ocean 0.4 km (0.25 mi) off of Palm Beach, Palm Beach County, Florida, in about 3.65 m (2 fathoms) of water on 8 May 2011. *Stephen M. Schelb*

Other common names: Hawksbill turtle, hawksbill, carey.

Similar species: The shell of the hawksbill differs from the other sea turtles in that its scutes overlap one another. These translucent plates are the source of "tortoiseshell." To harvest these valuable scutes, the turtles are killed; in some parts of the world, including Cuba, the take of this endangered species continues. Other than its scutes, this turtle is similar in general appearance to its other marine relatives.

General range of species: Widely distributed globally, the hawksbill turtle frequents subtropical and tropical oceans. It does not usually assemble in nesting concentrations but is more of a solitary species. It is dispersed through the tropics along mainland coasts, large oceanic islands, smaller islands, and tiny atolls with beaches that accommodate nesting sites.

Island distribution: This species is known on the barrier islands from only two live specimens (see below). Other than one cold-stunned specimen, stranded individuals have never been documented on the island beaches, nor were they included in early reports of incidental catch in the CHES fisheries relayed to LeBuff. It is undoubtedly rare that hawksbills pass through local waters. There are vague unpublished records of this species from a live-bottom site in the Gulf of Mexico very close to the beaches of northern Collier County. There is one reported, but undocumented, hawksbill nesting emergence on Longboat Key, Sarasota County, and a few strandings of this species in Collier County.

Preferred habitat type(s): In its preferred tropical habitats, *Eretmochelys* frequents coral reefs, rocky outcrops, and generally shallow waters with very productive live bottoms. Typically, the hawksbill sea turtle occupies waters less than 10 fathoms in depth.

Size: The hawksbill is not as small as perceived, based on the countless number of small specimens that were imported into the U.S. from the Caribbean basin as souvenirs or taxidermy-mounted specimens before the hawksbill was included under the ESA in 1970. Those that were imported pre-ESA are still legally hanging on many American home and office walls. The adult female hawksbill sea turtle can attain a weight of 127 kg (280 lb) and reach a straightline carapace length of 96.5 cm (38 in). At hatching, neonate hawksbill turtles have an average SCL measurement of 42.0 mm (1.65 in).

Color: The hawksbill is highly variable in coloration when young and is a beautiful creature. The carapace of an adult is generally dark greenish brown with radiating color mixes including reddish-brown, yellow, and orange tones. This turtle's plastron is typically yellow with a few randomly placed black streaks. Its carapace darkens with advanced age, and the rough scutes of the hawksbill can support epibionts that may equal the densities of those commonly found fouling the carapace of the loggerhead turtle.

Other characteristics: The specially adapted narrow head with the hawk-shaped beak and mouth parts, for which this turtle is named, is one major distinguishing feature of *Eretmochelys*.

Diet: From all sources, it appears that the hawksbill prefers sponges over all the other invertebrate organisms it consumes. The narrow beak allows *Eretmochelys* to reach into narrow recesses on coral reefs and extricate this favored prey. The hawksbill also consumes corals and hydroids, anemones, sea urchins, and a variety of mollusks in habitats where sponge populations are limited.

Reproduction: Where this species nests in the Caribbean, it does so between April and November, with a peak nesting frequency in June and July. Re-nesting occurs every two or three years. Little is known about the hawksbill's tendency for multiple nesting in any nesting population near Florida, but there is some evidence that the female deposits two to four clutches during a nesting year. The female produces an average clutch size of 161 eggs in the Caribbean. Incubation time, from deposition to hatchling emergence, is 50 to 70 days. The eggs of the hawksbill turtle may be the smallest among the sea turtles, with an average size of 38.0 mm (1.5 in).

This photograph of a hatchling *Eretmochelys imbricata* was taken on 29 August 2005 on Back Beach, Tobago, Republic of Trinidad and Tobago. Hatchlings of *Caretta*, *Eretmochelys*, and *Lepidochelys* are very similar in appearance, and knowledge of key characteristics is required to differentiate the genera at this stage. *Lauren Kirkland*

Life history: Life history characteristics are presumed to be similar to the other sea turtles. Like the others, post-emergent 19 g (0.67 oz) hatchlings immediately leave the beach in their swim frenzy mode, racing offshore to reach wherever it is that they seem to disappear. Once reaching their poorly identified (and not yet specifically located) developmental habitats, they go through a period of "invisibility." It is not until they are small, attractively marked subadults that they begin to show up in the usual hawksbill habitats.

Population status: The hawksbill population continues to decline worldwide. Overall, all size-classes of this overexploited species have been reduced over the past two centuries. Population estimates suggest that worldwide only 15,000 hawksbill turtles nest in any one year and that the species is globally endangered.

Threats: Throughout the world, this turtle faces serious survival problems despite efforts to curtail the harvest of tortoiseshell. Coastal development incessantly destroys the hawksbill's prime nesting habitat, and there is an increasing loss of its best foraging habitat because of expanding, pollution-caused, coral die-off. The interaction with fisheries because of incidental catch in trawls and gill nets is a major menace to all sea turtles, including the patchily distributed hawksbill populations. Egg harvesting for subsistence continues at a very high level in third-world countries.

Comments: In 1959 the hawksbill turtle was included on the original list of the amphibians and reptiles that occurred within the proclamation boundary of JNDDNWR because the species ranges along the coast of the region and is assumed to enter Pine Island Sound. Some of those waters are included within the refuge boundary. Documentation of sea turtle stranding events on island beaches started in 1959. Years later, beginning in 1977 and continuing for a decade, the senior author was authorized by USFWS to respond to reports of stranded sea turtles and provided documentation of those strandings to the National Sea Turtle Stranding and Salvage Network (STSSN). During his tenure as stranding coordinator for Sanibel and Captiva islands from 1959 until 1991, LeBuff never examined a hawksbill turtle from Sanibel or Captiva islands. Thus, the hawksbill was never confirmed to occur within the original refuge proclamation boundary.

On 21 June 2008, Ralph Woodring (pers. comm.) rescued a subadult hawksbill turtle that had become entangled in a piling of his boat dock along the Woodring Point shore of Pine Island Sound. A nearby fisherman had hooked the specimen, and the line parted, entangling the hawksbill with the piling and trapping it underwater. Unable to surface, the specimen was drowning. Woodring entered the water and caught the turtle, cut it loose, and applied a sea turtle resuscitation technique, which revived the hawksbill. The specimen was then transported to CROW, the Sanibel-based wildlife rehabilitation center, for observation. Thoroughly examined externally and internally by x-ray, the specimen was found to have no fish hooks imbedded in its throat or gut and was generally in good condition (P. Deitschel, pers. comm.). This hawksbill turtle was released near the point of capture on 23 June 2008.

Range Map 07. Pine Island Sound/Sanibel Island location of rescued *Eretmochelys imbricata*—2008.

In January 2010, water temperatures in this region's inshore marine systems plummeted coincidental to the passage of a severe cold front and the lingering low temperatures it left in its wake. These unusually low water temperatures to which marine organisms were exposed were recorded by a series of River, Estuary and Coastal Observing Network (RECON) sensors operated by SCCF since 2007. These sensors are strategically located at established stations in both the upper Caloosahatchee and selected sites in the CHES. Three of these data points are physically situated at the following locations: near Redfish Pass (Site A); near Blind Pass (Site B); and in the Gulf of Mexico near the entrance to San Carlos Bay (Site C), (Map 8). On January 11-12, 2010, these temperatures were, respectively: Site A, 8.73°C (47.7°F); Site B, 7.72°C (45.9°F); and Site C, 8.19°C (46.7°F). The threshold temperature below which severe hypothermic stunning and mortality of sea turtles occurs is 10°C (50°F) (Davenport 1997).

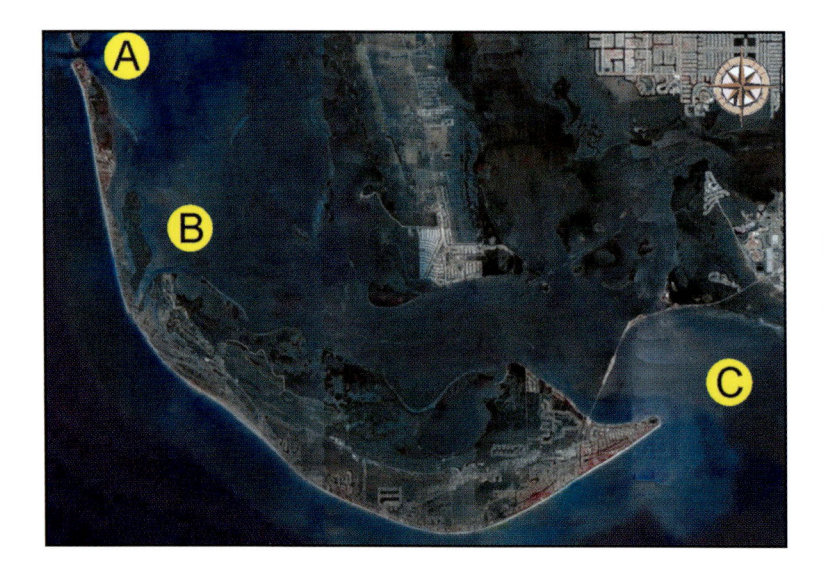

Map 8. The locations of three of the SCCF RECON sites from which water temperature data was collected are referenced as A, B, and C on page 103.

Twelve cold-stunned sea turtles were collected early on in the marine waters around Sanibel and Captiva islands during this frigid event. Of these, 11 were alive and responsive, and one was dead. One of the live specimens was identified as a hawksbill turtle by SCCF sea turtle coordinator Amanda Bryant (pers. comm.). Ten days later, after being sheltered and warmed, this specimen was PIT- and flipper-tagged and then released in the Gulf of Mexico at Vanderbilt Beach, Collier County. Ultimately, over the next days and weeks, the number of impacted turtles increased many-fold as carcasses of what were assumed to be hypothermic-stressed or advanced pneumonic-diseased sea turtles began to strand on the islands.

Gopherus polyphemus (Daudin, 1802)
Gopher tortoise

Young adult gopher tortoises usually reveal a sculptured appearance of the carapace. As these tortoises age, constant abrasion from daily passage through their sandy burrows wears the turtle's shell smooth. *Daniel Parker*

Other common names: Gopher.

Similar species: None in the eastern U.S.

General range of species: *Gopherus polyphemus* is a southeastern species found from extreme eastern Louisiana to southeastern South Carolina and as far south as Cape Sable, Florida.

Island distribution: Found throughout Sanibel and Captiva islands on high ground.

Preferred habitat type(s): *Gopherus polyphemus* is an upland species, found on high and dry ridges, as well as around developed uplands.

Size: Large, 23 to 38.7 cm (9 to a record maximum of 15.2 in). Soon after emergence from the egg when its shell has expanded and lost its egg shape, a hatchling averages 48.2 mm (1.9 in) in SCL.

Color: Primarily dark brown but can be gray. The plastrons are yellowish brown. Hatchlings are much lighter in color; they tend to be mostly yellow or orange with brown accents.

Other characteristics: The gopher tortoise has elephant-like limbs. The male has an enlarged gular scute at the front of the plastron that is used as a jousting weapon in combative squabbles during territorial disputes.

Diet: The gopher tortoise is primarily herbivorous, eating mostly grasses, but will also eat foliage from small shrubs, as well as flowers and fruit from various plants. It will sometimes eat animal material such as carrion, including dead fish from time to time, and is known to forage in seaweed on island shorelines. Sanibel Island specimens have been observed eating chicken scraps that were thrown around picnic sites. This tortoise can go for long periods without water and has the ability to store essential water for surviving the dry months. Several island residents with gopher tortoises on their property provide water for them during the dry season or even have trained them to drink from a hose. Of course, this is unnecessary to the gopher tortoise's survival during the dry season. It has evolved to cope without intervention.

Reproduction: The female *Gopherus polyphemus* begins breeding at about 23 cm (9 in), the male at about 18 cm (7 in). Depending on the latitude, this can be as early as nine years of age for females at the southern limits of this tortoise's range, or as late as 19 years near the northern limits. The female typically deposits between five and nine eggs in a single clutch. The eggs of the gopher tortoise become hard-shelled immediately after deposition and are nearly round.[17] The greatest diameter of a typical gopher tortoise egg averages 42.0 mm (1.65 in). Eggs are deposited in spring (May and June), and a clutch may be deposited in a shallow egg cavity the female digs in the mounded sand immediately outside the burrow. The eggs incubate for about 90 days before hatching.

A hatchling gopher tortoise. This specimen had recently left the egg cavity. The small white projection on the tip of this individual's nose is the caruncle, or temporary egg tooth, which most egg-laying reptiles develop and use to help them break through, or pip, their eggshell. Its presence on this specimen is evidence of its recent hatching: shortly after a neonate's emergence from the nest, the caruncle disconnects and falls away. *Charles LeBuff*

Life history: The name gopher tortoise derives from resident burrows it builds. The gopher tortoise is considered a keystone species in the Southeast. More than 400 different animal species throughout the range of the tortoise have been documented to use gopher tortoise burrows. The abandoned burrows of the adults are used by some juvenile tortoises that have not yet constructed their own burrows. The habitat of the gopher tortoise is dependent upon occasional fire to keep woody vegetation from becoming dominant. Active and inactive burrows serve as refuges for animals during a fire. Tortoises are not very social with each other, but tend to be found in clusters because they select the best habitats for their burrows.

The dome-shaped entrance of a gopher tortoise burrow conforms to the shape and dimensions of the resident turtle's shell. The burrow is continuously enlarged throughout its length as the tortoise grows. Gopher tortoise burrows slope downward and are usually straight with a chamber at the end to allow the turtle to turn around. They can reach a depth of as much as 3 m (10 ft) in some habitats. There are records of gopher tortoise burrows up to 12 m (40 ft) in overall length. Those on the barrier islands are known to be lengthy but not very deep because of the low elevation of the uplands and the high level of the ground water aquifer. Many other animals use the tortoise's burrow as a refuge or retreat during unfavorable weather or fire events. *Charles LeBuff*

Population status: The gopher tortoise population is doing quite well on Sanibel and Captiva islands. The restoration of conservation lands has been beneficial, and some island residents have helped address the needs of this tortoise. Some tortoises are actually relocated onto Sanibel from off-island projects. The gopher tortoise does very well in or near housing communities on Sanibel Island. The exception to this is the occasional tortoise that gets hit by a vehicle on the road or hatchlings that are killed by cats and dogs.

Threats: The gopher tortoise is state protected in all states in which it occurs. The USFWS lists the gopher tortoise as Threatened west of the Tombigbee River in Alabama. The FFWCC lists it as Threatened in Florida, and there is a petition to list the entire population as Threatened by USFWS. The biggest threat to the gopher tortoise is habitat loss as a result of upland development. Reptiles, birds, and mammals prey upon the gopher tortoise (mostly the juveniles). In some areas, it is eaten by local people, even though it is illegal to do so. Additional threats are discussed below.

Comments: In 1959 the gopher tortoise was abundant on Sanibel and Captiva islands, and colonies occupied all of the upland ridges. The tortoise was regularly taken by island families on a subsistence basis, and gopher and rice was a dish often consumed by a minority of the approximately 400 families that permanently resided on the islands. The tortoise population apparently could sustain this harvest, and its numbers remained robust until the 1970s. By this time, an exotic pest plant, the Brazilian pepper (*Schinus terebinthifolius*), had invaded the margins of the upland ridges and slowly encroached on the highest parts of these ridges. This fast-growing alien plant outcompeted the open grassy tortoise habitat, and prime tortoise habitat disappeared slowly, along with the tortoises. Vehicular traffic also took a toll, as all size classes of the gopher tortoise were killed crossing the heavily traveled island roads. Commuters from the mainland often stopped and picked up adult tortoises that were exposed while traversing roadways or foraging along the maintained road shoulders. Most were transported off-island and eaten.

In the 1950s and 1960s, the February festivities of the Edison Pageant of Light in Fort Myers always included a "Gopher Derby." Tortoises captured from throughout Florida, and even from southern Georgia, ended up as contestants in this popular tortoise race. After the derby many of these tortoises were released on Sanibel Island. Because of this practice, the genetics of the Sanibel tortoise population became muddled, and the resident tortoises were apparently infected by a serious pathogen previously unknown on the islands. In 1989 an infectious disease was reported among the Sanibel Island *Gopherus*, which was soon diagnosed as a tortoise mycoplasma, a detiologic agent identified as *Mycoplasma agassizii* (Brown et al. 1999) and given the name Upper Respiratory Tract Disease (URTD). According to McLaughlin (2000), 80 percent of the adult gopher tortoises on Sanibel Island were infected by URTD, and many infected tortoises succumbed to the ailment.

During the period when URTD was first reported, LeBuff confronted a group of non-resident landscapers and seized and rescued three large adult Sanibel Island gopher tortoises they had captured and had in their possession. These hapless tortoises were going to be taken to the mainland and would have ended up cooked with rice in a Crock Pot.

The specimens appeared symptomatic with URTD but in reality these "symptoms" were actually no different than specimens from within a small gopher tortoise colony that LeBuff had studied near Bonita Springs, Lee County, Florida, in 1953. Some of this mainland group contained individuals with foaming nares and an ocular discharge.

It is possible that URTD is nothing new among gopher tortoises in the region but is simply an issue that has been overlooked and only relatively recently recognized. Or, . . . whatever reason, URTD may have become more virulent to infected animals. Perhaps something, such as a commonly used pesticide, alters the normal immune response, leaving them more vulnerable.

Kinosternon baurii (Garman, 1891)
Striped mud turtle

The shell color of the adult striped mud turtle is variable in pigmentation. On this specimen the carapace stripes are well defined. *Daniel Parker*

Other common names: Mud turtle, musk turtle.

Similar species: Florida mud turtle.

General range of species: The striped mud turtle is a southeastern species ranging from the Florida Keys to as far north as coastal Virginia. This species is found mostly in the coastal plain of Georgia, the Carolinas, and Virginia.

Island distribution: Relatively common on Sanibel Island but found less frequently on Captiva. It thrives in the freshwater bodies of Sanibel's interior wetlands.

Preferred habitat type(s): *Kinosternon baurii* is found in freshwater bodies and in brackish waters with low salinities. They prefer still or slow-moving water but some populations occur in spring-fed rivers. It will aestivate or become dormant on land if water levels are low. It prefers areas with dense vegetation and soft bottoms.

Size: The adult female can reach 11.6 cm (4.6 in) in SCL and the male can be as large as 10.4 cm (4.1 in). The average size of neonates emerging from the egg is a mere 21.0 mm (0.8 in).

Color: Black or brown on the dorsal side and yellow on the ventral side. It has three orange or yellow longitudinal stripes down the length of its carapace, although some populations have no visible stripes. The head has two yellow longitudinal stripes on each side of its dark (mostly black) head.

Other characteristics: The striped mud turtle has plastral hinges that allow it to close its shell partially. It also has a raised keel down the middle of its back. Many older specimens become really dull or unicolored, losing the carapacial striping, and may be confused with the Florida mud turtle.

Diet: The striped mud turtle eats both plant and animal material, including carrion, snails, fish, algae, and aquatic plants.

Reproduction: *Kinosternon baurii* can deposit as many as three clutches of eggs per year. The average clutch is two or three eggs, but can be as large as six. This turtle's eggs have hard, calcified shells that are elliptical in shape. On average these measure 17.0 by 28.0 mm (0.68 by 1.1 in). The eggs are capable of undergoing a diapause mode, meaning that the embryological development can be stalled until the environmental conditions above the surface are adequate for egg success, such as at the beginning of the rainy season when embryo-sustaining soil moisture is available or at more suitable incubation temperatures.

This hatchling striped mud turtle has a nickel-sized shell and is typically colored for the young of this species. Like the adults, they are sometimes quite variable in coloration. This hatchling has the three distinctive stripes on its carapace. *Tim Walsh*

Life history: This small turtle can reach very high densities in good habitat and is frequently seen in ditches and canals. It is mostly nocturnal, especially during the warmer months.

When water levels are low, it will move overland from wetland to wetland and is often seen crossing roads.

Population status: The striped mud turtle is doing quite well on Sanibel Island. It occurs in fresh and brackish water bodies throughout the island. Captiva Island has very little habitat for this turtle, so its numbers are very low there.

Threats: The striped mud turtle is eaten by fish, reptiles, birds, and mammals in its natural environment. Sanibel Island has enough suitable habitat to sustain this species.

Comments: In 1959 the striped mud turtle was considered to be the most abundant turtle on Sanibel Island. It occurred everywhere freshwater existed and was often encountered at night crossing any of the five roads that passed through the interior freshwater wetlands. In the spring, when water levels were at their lowest, specimens of this turtle could be collected as they foraged in shallow, drying puddles in roadside ditches. During periods of high water, nickel-sized hatchlings were occasionally collected as they crossed Sanibel-Captiva Road, which follows the mid-island ridge on Sanibel Island. A specimen collected DOR on Sanibel in June 1959 was furnished to OSU and recorded as R-1723.

Kinosternon subrubrum steindachneri (Siebenrock, 1906)
Florida mud turtle

An adult Florida mud turtle from Polk County, Florida. This species is becoming jeopardized throughout most of Florida. *Daniel Parker*

Other common names: Mud turtle, musk turtle.

Similar species: Striped mud turtle.

General range of species: *Kinosternon subrubrum* is mostly a southeastern species ranging from eastern Texas to the eastern coast of the U.S. The Florida mud turtle is found throughout the Florida peninsula and as far north as New Jersey in the east and northern Indiana to the west. The subspecies *K. s. steindachneri,* is found from northern peninsula Florida to the southern tip of the peninsula, but is absent in the Florida Keys.

Island distribution: Until recently, the Florida mud turtle was considered extirpated on Sanibel and Captiva islands. On 10 August 2012, a hatchling was collected and photographed near Wulfert Point (Sanctuary Golf Course), Sanibel Island, by Kyle Sweet, superintendent of the golf course. Sweet found a second hatchling nearby on a golf course green on 17 August.

Preferred habitat type(s): The Florida mud turtle prefers shallow, soft-bottomed, fresh or somewhat brackish water bodies. This turtle seems to thrive in marshes, as well as ditches and canals, and even occupies slow-moving creeks.

Size: The adult Florida mud turtle is usually 7.9 to 11.9 cm (3.1 to 4.7 in) in SCL. The average SCL of *K. s. steindachneri* hatchlings is 23.0 mm (0.9 in).

Color: Mostly brown or black on the carapace, with a yellowish plastron. The head and limbs are also brown or black with no stripes, although sometimes these appendages have light brown mottling. Hatchlings often have a bright red to yellow plastron.

Other characteristics: The Florida mud turtle has plastral hinges that allow it to close its shell. This turtle has a hook-beaked upper jaw. Some specimens have a "clawed" tail—that is, the tip of the tail is bisected and noticeably hardened. The reason for this adaptation is unknown. The subspecies, *steindachneri*, is prone to bite readily when picked up. Nearly all the specimens that Lechowicz has handled have opened their mouths and attempted to bite. This has also been the case with the subspecies *subrubrum* of the north, as well as some *K. bauri*.

Diet: The Florida mud turtle consumes both plant and animal material. It eats crustaceans and other hard-shelled prey, as well as various plant materials.

Reproduction: *Kinosternon subrubrum* can deposit three or more clutches of eggs in a season, usually with two to four eggs in each clutch. The incubation time extends up to 100 days, as it does for *K. baurii*. Eggs may undergo embryonic diapause in which the eggs cease development for a period of time, resuming development under favorable environmental conditions. The eggs are elliptical and calcified, with a granular surface. The average diameters are 15.0 by 28.0 mm (0.6 by 1.1 in).

Life history: This largely aquatic species is often overlooked due to its secretive nature, dark appearance, and small size. It is fond of heavily vegetated canals and ditches in which it scurries along the bottom looking for prey. During low water it migrates overland and even aestivates in the ground until the rains begin and ponds and ditches fill. It was once very common throughout Florida, but herpetologists have noted a significant decrease in its population in recent years.

A hatchling Florida mud turtle collected at the Ordway-Swisher Biological Station, Putnam County, Florida. *C. Kenneth Dodd, Jr.*

Population status: The Florida mud turtle was known from only a few adult specimens collected and identified by LeBuff on Sanibel Island in the 1960s and early 1970s. The species was rediscovered in 2012 when two hatchlings were found on western Sanibel Island at the Sanctuary Golf Course.

Threats: The Florida mud turtle is eaten by various fish, reptiles, birds, and mammals. Its recent scarcity throughout Florida has resulted in an increased demand from the pet trade. Roads are a problem for this animal since it inhabits roadside ditches.

Comments: After the Florida mud turtle was first collected in the BT of JNDDNWR in 1960, this species was frequently observed in the same habitats occupied by its close relative, the striped mud turtle, *K. baurii*, in the central wetlands of Sanibel Island. The Florida mud turtle was never considered to be common on the island. It was occasionally encountered crossing roads when the wetlands were flooded, at least until the mid-1970s. According to Lechowicz (2009), this turtle was considered rare on Sanibel Island in 2009 and until recently was considered extirpated.

Lepidochelys kempii (Garman, 1880)
Kemp's ridley sea turtle

An adult female *Lepidochelys kempii* that has just finished nesting on Sanibel Island near Bowman's Beach on 22 April 2011, and is returning to the Gulf of Mexico. *Amanda Bryant*

Other common names: Atlantic ridley, Gulf ridley, bastard turtle.

Similar species: There are two distinctly different ridleys in the Atlantic Ocean. One is the species under discussion; the other is the Pacific or olive ridley (*Lepidochelys olivacea*). Despite its accepted common name, the latter is not restricted to the Pacific Ocean. This wide-ranging sea turtle also occurs in the Western Atlantic and is found from the northern coast of the South American continent into the Caribbean, around Jamaica and occasionally as far as Cuban waters. Both live and dead stranded individuals were first documented to occur in extreme South Florida waters in 1999. Over time, the Pacific ridley may reach the Gulf of Mexico in numbers where it may be confused with *Lepidochelys kempii*. It could also eventually interbreed with the latter species. Unless the remaining Kemp's ridley stock rapidly increases from its currently endangered level and regains its robust 1940s population, interbreeding could ultimately lead to genetic changes in this species as we know it today.

General range of species: Kemp's ridley is distributed in the northwestern Atlantic Ocean, and primarily in the Gulf of Mexico. Individuals are commonly found in New England waters and Long Island Sound, and the species has been recorded from northern Europe and the Mediterranean Sea, carried by currents across the Atlantic.

Island distribution: Occurs in all waters of the region; the female very rarely nests on the barrier islands.

Preferred habitat type(s): In the vicinity of Sanibel and Captiva islands, Kemp's ridley frequents the shallows of the CHES and live bottoms offshore in the Gulf of Mexico.

Size: *Lepidochelys kempii* is the smallest among sea turtles. Hatchlings have an SCL of 38.1 mm (1.5 in) and weigh 17.1 g (0.6 oz). One of the largest adult specimens ever measured was found stranded on Point Ybel, Sanibel Island, in 1971. This specimen[18] had a straight-line carapace length of 70 cm (27.5 in). A large adult averages 45 kg (100 lb).

Color: The carapace and dorsal sides of the body of this sea turtle are uniformly a grayish-green color. The plastron is a pale tone of yellow.

Other characteristics: The carapace of Kemp's ridley is unique in that the shell can sometimes be wider than it is long. This turtle's beak is almost parrotlike in profile, so the appearance of its head is much different from that of other sea turtles. This species is also unique in that it usually comes ashore and nests during daylight; other sea turtles (except *Lepidochelys olivacea*) are night-time nesters, although there are rare cases where individuals of other species have made daylight landings.

Diet: Kemp's ridley preys on a variety of bottom-dwelling invertebrates, including crabs and mollusks. This species also consumes jellyfish.

Reproduction: The egg size of the Kemp's ridley averages 39.0 mm (1.53 in) in diameter. The female typically comes ashore to nest during daylight hours and often on windy days with strong onshore winds. Where this species assembles for nesting in Mexico, it has historically formed what is known as an *arribada* (literally meaning an arrival). Even today, an arribada can total hundreds of females who come ashore all at once in a wave, almost as one united group.

This hatchling *Lepidochelys kempii* was photographed on 2 July 2010, on South Padre Island, Texas. *Aaron Garstin*

Life history: Prior to discovery of the Mexican nesting site of Kemp's ridley, it was a popular belief among those engaged in the now-defunct Florida commercial sea turtle industry that the ridley was a hybrid of loggerhead and green sea turtles. Because of this misconception, it was commonly known as the bastard turtle.

A film, shot in 1947 by engineer Andres Herrera while flying in a light plane along the Mexican gulf coast near Rancho Nuevo, Tamaulipas, was not disclosed to the scientific community until 1963. Herrera stumbled across the incredible daylight mass nesting of an estimated 40,000 Kemp's ridleys and filmed the event. The plane landed on the beach, and this spectacular assemblage was captured on film for posterity. This was one of the most amazing biological discoveries of the 20th century.

Population status: After the 1940s, the Kemp's ridley population went into a steep decline. Increasing shrimping operations in the primary non-nesting population center in the northern Gulf of Mexico, coupled with the pressures of unrestricted egg harvesting on the Mexican nesting beach, were responsible. The lowest level of nesting occurred in 1985 when only 702 female Kemp's ridleys were recorded on Rancho Nuevo beach. Since then, with protection provided by the ESA and the Mexican government, the implementation of TED requirements, and tougher enforcement of conservation regulations by Mexican officials on the nesting beach, the crisis has somewhat diminished. As a result, the survival of this unique species has become more hopeful. In 2012, 9,000 female Kemp's ridleys nested in Mexico (2013 nesting totals are unavailable at this writing), and in 2013 153 did so in Texas, mostly on the beaches of Padre Island National Seashore. There is now a glimmer of hope for long-term survival of the species.

Threats: Commercial fishing remains the greatest threat to this species for the foreseeable future. Destruction of this ridley's prime habitat, which includes degradation of its limited nesting beaches caused by human development, continues. The ruin of live bottoms, which are increasingly impacted by trawling and deepwater petroleum exploration and production,[19] will unrelentingly affect the long-term survival of this species and its marine relatives negatively.

Comments: Only stranded loggerheads were tabulated on Sanibel and Captiva islands between 1959 and 1971. On 21 December 1971, George Weymouth, at the time a temporary JNDDNWR staff member, discovered the carcass of a large Kemp's ridley on the Gulf beach at Point Ybel. LeBuff (1990) provided data on this specimen's skull size and straight-line carapace (SCL) dimensions (see above under Size). Following this stranding, additional strandings of this species became more commonplace on Sanibel and Captiva islands, reaching their greatest numbers in the late 1980s.

Prior to the 1971 Kemp's ridley stranding, this species was known to occur regularly in moderate numbers in the waters of the CHES. This was frequently verified when commercial gill-net fishermen unintentionally caught subadults of this species. In many cases these specimens were held and turned over to the senior author for tagging, documentation, release, and in rare instances for rehabilitation.

In 1996, the unexpected happened when a female Kemp's ridley made a daytime landing on the western Sanibel Island beach. A beachfront homeowner who witnessed this event photographed the turtle, and the marine turtle unit of SCCF monitored the success of this nest. Five dead neonates were discovered when the nest was excavated, and the remaining contents of the nest cavity were examined post emergence. These were deposited in the collection of the FLMNH (Catalog numbers 10556-70), (K. Krysko, pers. comm.). Near midday on 22 April 2011, another nesting Kemp's ridley was observed on the gulf beach of western Sanibel.

Range Map 08. Site of the first known Sanibel Island nesting landing by a *Lepidochelys kempii* in 1996.

An important but heretofore unpublished fact related to sea turtle nesting on Sanibel and Captiva islands is that between 1959 and 1971, nighttime patrols to monitor nesting activity and other aspects of the sea turtle conservation program were not usually launched until the first week of May. Patrol startups were predicated on observations of turtle activity made during frequent aerial reconnaissance flights conducted by refuge staff in the JNDDNWR aircraft. Following patrols over Pine Island Sound to the various satellite refuges, which are administered from Sanibel Island, the return flights to the seaplane's base on Sanibel were routinely flown along the barrier island chain of beaches. This was done to spot any turtle crawls on the beaches below or any turtles immediately offshore. The sighting of a crawl would mark the beginning of sea turtle nesting season, and the nighttime work would commence after nesting activity actually was sighted on the beach.

The Kemp's ridley typically begins nesting earlier in the year than the loggerhead. The shallow tracks that the nesting female of the lightweight Kemp's ridley leave on the beaches are less conspicuous than those left by the much heavier and more abundant loggerhead. The Kemp's ridley does not leave noticeable, long-lasting flipper imprint scars on the beach, and the emergence tracks created by this small turtle crossing the beach are very short lived. Wind and precipitation soon obliterate them, and without evidence on the dry beach they can easily go unnoticed and undocumented.

We are convinced that the Kemp's ridley has regularly, albeit rarely, nested on the Sanibel Island beach prior to the first documented nesting visit in 1996. Because of the timing of human schedules, however, early nesters were likely unnoticed or overlooked on these barrier islands. A Kemp's ridley was witnessed nesting on Sanibel Island and remained on the beach from 1115 to 1230 hours on the morning of 22 April 2011 (see above photograph). It was the first sea turtle nest of the 2011 season.

Malaclemys terrapin macrospilota (Hay, 1904)
Ornate diamond-backed terrapin

The light-colored centers of the carapacial scutes of the above adult female ornate diamond-backed terrapin make identification of this subspecies easy. *Chris Lechowicz*

Other common names: Terrapin.

Similar species: None.

General range of species: Found along the east coast of the U.S. from Cape Cod, Massachusetts, to the gulf coast of Texas. Individuals have been found recently in Bermuda, where it is believed to be native. The form occurring on Sanibel and Captiva islands is *Malaclemys terrapin macrospilota*, which ranges from Florida Bay to the western panhandle of Florida. It is one of seven subspecies.

Island distribution: Found in mangrove areas such as tidal creeks and pools with moderate salinity.

Preferred habitat type(s): *Malaclemys t. macrospilota* is a turtle of brackish water. The terrapin is typically associated with mangrove swamps throughout its range but also occurs in salt marshes, as well as the mangrove forests along the west coast of Florida. The preferred habitat of other members of the genus *Malaclemys* consists of salt marshes as opposed to mangrove swamps.

Size: The adult male *Malaclemys t. macrospilota* attains an average of 12.49 cm (4.9 in) in carapace length and the female averages 18.06 cm (7.1 in). The maximum SCL for this subspecies was obtained from a female that measured 23.8 cm (9.375 in). The average SCL of hatchlings of this subspecies is 32.0 mm (1.26 in).

Color: The carapace is black to gray with orange to yellow blotches in the center of each pleural scute of the upper shell. The plastron varies from black to orange. The head is white to gray, with or without black spots. The limbs are typically the same color as the head and have similar black spots.

Other characteristics: The ornate diamond-backed terrapin has well-developed webbed feet for swimming. It typically has knobs or a median keel running down the vertebral scutes. The knobs are extremely obvious in hatchlings and juveniles.

Diet: It eats snails, bivalves, crabs, fish, insects, and vegetation but has a preference for the common periwinkle snail (*Littorina littorea*).

Reproduction: *Malaclemys t. macrospilota* deposits an average of 5.75 eggs in a clutch and has been reported to deposit one to three clutches in a season. When they are first deposited, the eggs of *Malaclemys* are somewhat similar to those of the sea turtles in that they are leathery and flaccid. This looseness creates an obvious dimple in the egg's integument. As in the eggs of sea turtles, the surface depressions disappear and the shell becomes turgid as development of the egg begins and uptake of ambient moisture from the nest cavity increases pressure within the egg capsule. Unlike the nearly perfectly round eggs of sea turtles, the eggs of the diamond-backed terrapin are elliptical, their width and length averaging 20.0 by 30.0 mm (0.78 by 1.18 in) in diameter.

A young ornate diamond-backed terrapin. The strange, conspicuous, black knobs or rounded keels along the dorsal portion of this post-hatchling's carapace show up well in this photograph. These tubercles are normal and may be nearly lost with advanced age; their prominence mostly disappears in females, but this keel-like feature can be retained into old age by the male of this subspecies.
Maggie May Rogers and Chris Lechowicz

No terrapin nests have ever been found on Sanibel or Captiva islands, although adults and hatchlings are occasionally seen. Some terrapins may nest on smaller natural mangrove islands with tiny but sufficiently elevated beaches. A series of dredged spoil islands in Pine Island Sound near Matlacha Pass also may be used as nesting sites.

Life history: This estuarine turtle lives in water bodies with variable, fluctuating salinities. In times of high salinity, it will often get its required freshwater directly from the surface of the water during rain events. The juvenile terrapin has been reported to spend time on land in moist areas, but the adult is mostly aquatic, leaving the water only to deposit eggs or to bask on floating logs and other structures. The terrapin has been described as being a freshwater turtle that lives in a marine environment.

Population status: *Malaclemys t. macrospilota* is somewhat common on the islands but is rarely seen because of the difficulty of traversing its habitat and the well-developed wariness inherent in its behavior.

Threats: In nature, the terrapin is eaten by fish, birds, and mammals. It is collected for human consumption in some areas but not traditionally in Southwest Florida. It is also collected for the pet trade throughout its range. The biggest threat to its population comes from crab pots, which prevent by-captured terrapin from getting air from the surface. Many abandoned crab pots continue to capture terrapins, leading to their death and therefore attracting more terrapins. Legislation in a few states requires a bycatch reduction device (BRD) to be installed in every licensed crab pot. The BRD prevents larger terrapins (mostly females) from entering the traps.

Comments: The ornate diamond-backed terrapin is known from locations throughout the red and black mangrove forests of Sanibel and Captiva islands. The creation of the mosquito control dike on JNDDNWR provided what should be exceptional nesting habitat for this species where it was completely nonexistent before. Adult females are regularly encountered on this dike, now known as Wildlife Drive, but a terrapin in the act of nesting has never been reported. Hatchlings were frequently collected from temporary rainwater puddles on the drive in the years prior to its paving and on unsurfaced Woodring Point Road. LeBuff once collected a hatchling terrapin on the gulf beach 0.75 km (0.47 mi) west of the SILS.

In about 1993, Michael Boerema, then an interpretive naturalist at the Tarpon Bay concession operation on JNDDNWR, showed the senior author a photograph that had been taken on JNDDNWR along the Commodore Creek Canoe Trail. It was an image of an unusually colored terrapin that had been photographed while the specimen was partially exposed basking on a section of overhanging red mangrove limb. Based on the turtle's much different coloration from the typical resident *Malaclemys* and the presence of a definitive neck pattern, LeBuff cautiously identified the individual as an adult *Malaclemys terrapin rhizophorarum*.[20]

Malaclemys t. rhizophorarum is currently considered to be confined to the more remote islands of the Florida Keys. Ernst et al. (1994) restrict the range of this subspecies to the Keys, as does Bartlett and Bartlett (2006). Much earlier, Johnson (1952) placed the northern periphery of this subspecies at Naples, Florida, where it would encroach well into the range of *M. t. macrospilota*. Based on this interesting observation, it appears that terrapins that externally resemble this more southern subspecies may occasionally range as far north as Sanibel Island. More detailed investigations relative to the diamond-backed terrapin population of the CHES are necessary in order to understand the distribution patterns of the diamond-backed terrapin along Florida's southwestern coast.

Although the mangrove system associated with the estuarine complex is large, the availability of nesting habitat for the diamond-backed terrapin remains limited. Lechowicz, as part of the SCCF Diamondback Terrapin Project, is currently conducting research on terrapins on and near Sanibel. He is documenting seasonal movements, courtship behavior, nesting locations and population assessments with mark-recapture and satellite telemetry.

In a study of this subspecies farther north in Florida's Citrus County, the blue-crab trapping industry was found to be a serious threat to the diamond-backed terrapin (Boykin 2005). Terrapins enter the submerged traps (known in the crabbing fishery as pots) seeking the bait, and without any modification to the traps to create an opening in the trap walls or top that would permit their escape, the turtles drown. Abandoned crab traps are a constant attractant, and they doom many terrapins.

More recently, Butler and Heinrich (2007) trapped diamond-backed terrapins along the Florida gulf and Atlantic coasts at eight study sites. This work was conducted to evaluate the efficacy of an experimental BRD developed for installation in blue-crab pots in an effort to reduce diamond-backed terrapin mortality. The rectangular experimental device, fabricated from heavy wire and measuring 4.5 by 12 cm (1.8 by 4.7 in), is wired into each of the four funnel-like openings located on each side around the perimeter of a standard square blue-crab pot. This device was designed to reduce the size of the trap's openings and was found to prevent the passage of the adult female and larger male diamond-backed terrapin into the trap body to reach the crab- and turtle-enticing carrion bait. Harvestable, legal-size blue crabs could still pass through the BRD openings and get into the pot. Butler and Heinrich found that 73.2 percent of the terrapins caught in crab pots could have been prevented from entering the traps had BRDs been installed. Based on these results, the authors recommended that the FFWCC adopt a rule that BRDs, equal in dimension to that of their design, be required in the commercial and recreational blue-crab industry in Florida. Many of the other coastal states with declining populations of diamond-backed terrapins, and viable blue-crab fisheries, already require the use of similar BRDs. As of 2013, Florida still had not adopted a comparable rule. Far too many diamond-backed terrapins continue to die because of their interaction with crab pots in Florida and state inaction.

Through the 1970s and 1980s, LeBuff frequently inspected abandoned or unattended blue-crab pots around the national wildlife refuge islands in Pine Island Sound, Matlacha Pass (both in Lee County), and Turtle Bay (Charlotte County). He often discovered the remains of terrapins inside the wire traps and occasionally was able to rescue and liberate live turtles.

Pseudemys nelsoni Carr, 1938
Florida red-bellied cooter

An adult Florida red-bellied cooter rests as it floats at the water's surface. Both hatchlings and adults are easily confused with other freshwater species. Characteristics of the front of the red-bellied cooter's mouth are much different from the mouthparts of similar turtles and make identification easy for someone who knows the distinguishing characteristics. (See text). *Bill Love*

Other common names: Florida redbelly turtle, red-belly cooter, cooter, striped-neck terrapin, streaky-neck turtle.

Similar species: Peninsula cooter, Florida chicken turtle, yellowbelly slider.

General range of species: Found in Florida and extreme southeastern Georgia, from Florida Bay in the south to the Florida panhandle and just north of Jacksonville, Florida, in Georgia. The range extends to the Apalachicola River basin in the Florida panhandle.

Island distribution: Found throughout Sanibel Island in all freshwater bodies.

Preferred habitat type(s): *Pseudemys nelsoni* inhabits still or slow-moving water bodies, such as lakes, ponds, streams, slow-moving rivers, canals, and ditches. This cooter prefers permanent water with dense aquatic vegetation.

Size: *Pseudemys nelsoni* is a large, basking freshwater turtle. The male typically averages 16 to 28.7 cm (6.2 to 11.2 in) in carapace length, and the female averages 28 to 33.3 cm (11 to 13.1 in) in SCL. The maximum SCL for this turtle is 37.5 cm (14.5 in). The average SCL of hatchling *P. nelsoni* is 30.0 mm (1.18 in).

Color: The carapace of *P. nelsoni* is typically black or brown with one (red to yellow) vertical stripe on every pleural scute. The stripe on the third pleural scute is wider than the rest and is usually reddish in pigmentation. The plastron is red to yellow, usually with no blotches, although older individuals show melanism and appear to develop black blotches. The head, limbs, and tail are black with yellow stripes.

Other characteristics: *Pseudemys nelsoni* has fewer neck stripes than *P. peninsularis*. The Florida red-bellied cooter has two tooth-like cusps on the upper jaw and one on the lower jaw. Similar freshwater turtle species lack these cusps.

Diet: The Florida red-bellied cooter is mostly herbivorous, eating a wide variety of aquatic vegetation. The juvenile is more carnivorous but still eats vegetation.

Reproduction: *Pseudemys nelsoni* sometimes deposits eggs in American alligator (*Alligator mississippiensis*) nest mounds. In this way it uses the female alligator to protect its nest from predators such as raccoons. Of course, the turtle has to deposit its eggs quickly while the female alligator is out hunting and not protecting her own nest. The Florida red-bellied cooter lays three to six clutches of seven to 26 eggs a year (averaging 14.3 eggs per clutch). The well-calcified but pliable, elliptical eggs measure 24.0 by 36.5 mm (0.94 by 1.44 in).

This colorful hatchling Florida red-bellied cooter can be easily confused with the neonates of the peninsula cooter, the sliders, or chicken turtle. *Daniel Parker*

Life history: The adult Florida red-bellied cooter often has tooth marks and scratches from attempted alligator predation. An alligator of moderate size is not usually able to crush the shell of an adult Florida red-bellied cooter because of the unusually thick bones making up the shell. *Pseudemys nelsoni* occurs sympatrically with *P. peninsularis*, *P. floridana*, and *P. concinna* in different parts of its range.

Population status: *Pseudemys nelsoni* is not nearly as common as *P. peninsularis* on Sanibel Island.

Threats: *Pseudemys nelsoni* is eaten by fish, birds, and mammals, including people in certain areas. Human consumption of other less common cooter species (e.g., *P. c. suwanniensis*) has led to the protection of *P. nelsoni* in Florida because of its physical similarity.

Comments: The Florida red-bellied cooter was never considered abundant on Sanibel Island. In earlier years, the island's red-bellied turtle population was patchily distributed. Once permanent freshwater came to the island as a result of mosquito control operations, individuals dispersed island-wide. Increased frequency of observations of *P. nelsoni* suggest this species has become more successful in the years following the alterations to the interior wetland basin. In May 2012, a Florida red-bellied cooter was seen by Joel Caouette (SCCF biologist) in the beach surf near the Gulfside City Park on Sanibel Island.

Pseudemys peninsularis Carr, 1938
Peninsula cooter

The peninsular cooter is extremely abundant on Sanibel Island. The insular population grew rapidly between 1959 and 1980, commensurate with the availability of permanent freshwater in the form of mosquito ditches and subdivision real estate ponds. *Daniel Parker*

Other common names: Florida cooter, cooter, striped-neck terrapin, streaky-neck turtle.

Similar species: Florida red-bellied cooter, Florida chicken turtle, yellow-bellied slider.

General range of species: The peninsula cooter is an endemic species of Florida. It is found south of Alachua County (Gainesville), Florida, to the Florida Keys.

Island distribution: Found throughout Sanibel Island in freshwater bodies.

Preferred habitat type(s): *Pseudemys peninsularis* lives in still or slow-moving water bodies, such as lakes, ponds, streams, rivers, canals, and ditches. It prefers areas with dense aquatic vegetation.

Size: *Pseudemys peninsularis* is a large aquatic turtle with a highly domed shell. The adult peninsula cooter averages 25 to 40 cm (25 to 15.7 in) in SCL. The adult male typically averages 13 to 32 cm (5.1 to 12.6 in) in carapace length; the female averages 26 to 38 cm (10.2 to 14.9 in) in SCL. The largest recorded length for this species reached 40.3 cm (15.875 in) SCL. The average SCL of hatchling peninsula cooters is 36.0 mm (1.4 in).

Color: The carapace of *P. peninsularis* is typically black or brown with thin yellow vertical lines on the scutes. The plastron is yellow and without blotches. The head, limbs, and tail are black with yellow stripes.

Other characteristics: The top of the head has two U-shaped lines that resemble hairpins. The jawline is not cusped, but straight. The vertical line/lines on the third pleural scute are wider than the other vertical lines on other scutes, but are not as wide as those on the Florida red-bellied cooter.

Diet: The peninsula cooter is mostly herbivorous, eating a wide variety of aquatic vegetation. The juvenile is more carnivorous but also consumes vegetation.

A hatchling peninsular cooter. By studying the hatchling images and the descriptions in the individual species' text, someone interested in identification can soon learn to tell the differences between this turtle and the hatchlings of the Florida red-bellied cooter, chicken turtle, and sliders.
Daniel Parker

Reproduction: *Pseudemys peninsularis* typically digs three nest chambers, consisting of a central cavity, which contains the most eggs, and two peripheral chambers, which may contain a few eggs. Total clutch size is 11 to 16 eggs. A female can deposit up to six clutches in a year. The eggs of this cooter are somewhat calcified but are still pliable and elongated. Their width and length average 26.0 by 34.0 mm (1.02 by 1.34 in) in diameter.

Life history: The peninsula cooter is a common turtle throughout its range. It is the basking turtle that is most often seen in South Florida. The peninsula cooter is often difficult to differentiate from the Florida red-bellied cooter while basking, especially when covered with duckweed and muck.

Population status: *Pseudemys peninsularis* is likely the most common turtle species on Sanibel Island, rivaled only by *Apalone ferox*. The peninsula cooter is commonly seen crossing roads, basking, and nesting on Sanibel Island. The hatchlings are often observed while seeking water bodies after leaving the nest.

Threats: *Pseudemys peninsularis* is eaten by fish, birds, and mammals, including people in certain areas. Human consumption of less-common cooter species (*P. c. suwanniensis*) has led to the protection of *P. peninsularis* in Florida because of its physical similarity.

Comments: The first two specimens of the peninsula cooter collected by LeBuff on Sanibel Island were likely new recruits to the herpetofaunal community of the island. Both were found in 1959. The first was encountered in the surf at Point Ybel, and the second was moving across Woodring Point Road from Pine Island Sound into the red mangrove forest fringing the shore. It is assumed that both individuals were transported down the Caloosahatchee and were fortunate to have reached Sanibel Island. Over time, this species became established and now commonly occurs throughout the freshwater interior of Sanibel Island, supported by habitat provided by mosquito control ditches and some real estate borrow ponds. Basking surveys by Lechowicz show that the peninsula cooter is the most common basking turtle on the Sanibel River and most other large water bodies. A few ponds have higher concentrations of basking sliders (*Trachemys*).

Terrapene carolina bauri Taylor, 1894
Florida box turtle

Pictured above is a handsome adult Florida box turtle. The base color of the carapace is highly variable and can range from a light tan hue to jet black. The yellow striping on the shell can often be lighter and broader in width than that of the above example. *Daniel Parker*

Other common names: Box turtle, box tortoise.

Similar species: Ornate box turtle (not found in Florida).

General range of species: The Florida box turtle is currently considered a subspecies of the eastern box turtle, *Terrapene carolina*. Not all herpetologists are in agreement because some molecular data suggest the Florida box turtle should be elevated to full specific status (C. K. Dodd, Jr., pers. comm.). *Terrapene c. bauri* is found from extreme southeastern Georgia south through the Florida peninsula, and reaches the Florida Keys.

Island distribution: Found throughout Sanibel and Captiva islands with the exception of the mangrove forest.

Preferred habitat type(s): *Terrapene c. bauri* is found in both mesic (lowlands) and xeric (uplands) habitats in Florida. It is usually associated with woodlands but also inhabits dry grasslands. The box turtle is usually thought of as preferring moist wooded areas covered with leaf litter and high humidity, but it also lives in significantly drier uplands in South

Florida. On Sanibel Island, it is often found in the vegetated areas of the Gulf Beach Ridge Zone.

Size: At hatching, the neonate Florida box turtle is tiny, averaging 33.0 mm (1.3 in) SCL. Adults on Sanibel Island range between 8.0 cm and 18.7 cm (3.2 to 7.4 in) SCL. The Sanibel individual that reached 18.7 cm is the maximum recorded SCL for *Terrapene c. bauri* in Florida.

Color: The Florida box turtle is primarily black to brown on its carapace, with yellow streaks. Some individuals are exquisitely colored dorsally and have a jet-black carapace with bright yellow stripes of delicate pinstriping. Generally, the plastron is yellow with occasional dark patches. The head, limbs, and tail are also dark with yellow markings.

Other characteristics: All members of the genus *Terrapene* are able to completely close their shell, so that the head, limbs, and tail are fully protected. On the male the plastron has a concave area that aids in copulation. Many box turtles on Sanibel Island show pronounced flaring of the rear marginal scutes. There seems to be a high percentage of very large box turtles in this population. Many males have bluish highlights on the face.

Diet: The Florida box turtle is an opportunistic feeder. It eats plant and animal matter, as well as fungi. It consumes carrion and will enter water to feed on aquatic prey. Its favorite food appears to be earthworms.

Reproduction: *Terrapene c. bauri* breeds throughout the year in Southwest Florida. This turtle has been reported to produce one to three clutches of one to five eggs (2.7 average) in a season. The eggs are pliable, oblong, and on average measure 20.5 by 37.0 mm (0.8 by 1.45 in).

A very young, possibly a few weeks old, post-hatchling Florida box turtle. The young are rarely seen and very little is known about their early life histories. *Bill Love*

Life history: The juvenile box turtle is seldom encountered in the field; it is a master of the art of concealment, and little is understood of the species' early life history. The Florida box turtle is a small, long-lived turtle. Its survival depends upon the long reproductive life of the adult, as it has few offspring each year. This species is also known to have a small home range, but that of some individuals may be similar to other box turtles. Its secretive nature often allows it to go unnoticed in many areas for years. It is considered terrestrial, but a better characterization would be semi-aquatic since it spends time in shallow water bodies, especially after rainfall. It will enter marshes, lakes, and streams from time to time, and on rare occasions the Gulf of Mexico (Photo documentation submitted to Lechowicz, August 2012.)

Population status: The Florida box turtle is common in some areas on Sanibel and Captiva islands, but box turtle numbers have decreased significantly since the writings of George Campbell in the 1970s and 1980s.

Threats: The box turtle is most threatened by habitat loss, road mortality, and predation, and by people picking individual turtles up as pets, even though this practice is now prohibited. Habitat loss is a serious problem in many states, although this is not considered a problem on Sanibel and Captiva islands. The box turtle often crosses roads in the spring after rain during the day. This leads to frequent road mortality, even on Sanibel and Captiva islands. Raccoons and fire ants seem to be the most serious predators on box turtles in South Florida; both find nests and may eat eggs or pipped hatchlings. Both the native and imported species of fire ants (genus *Solenopsis*) prey on juveniles or diseased adults that are not very mobile. Raccoons can prey on all size-classes of box turtles. Fire, both wild and controlled, takes a heavy toll on box turtles. Fire-scarred specimens that have survived earlier serious burns are common in all populations of the box turtle.

Comments: In 1959 the Florida box turtle was abundant on Sanibel and Captiva islands and remained so until the late 1960s. It was not uncommon to observe two or three live specimens any morning as they tried to negotiate a crossing of Periwinkle Way or Sanibel-Captiva Road. Since then, this species has become somewhat rare in areas where it once was abundant.

Two DOR specimens of Sanibel Island box turtles were supplied to the collection of OSU, in 1959. These are cataloged as R-1667 and R-1729.

A mark-recapture study of this species on the islands has been in effect for several years. More than 100 animals have been processed and released. The Florida state record length for *T. c. bauri* is actually from Sanibel Island at 18.7 cm (7.36 in) SCL.

Trachemys scripta scripta (Schoepff, 1792)
Yellow-bellied slider (Actively introduced)

An algae-covered yellow-bellied slider. *Bill Love*

Other common names: Slider.

Similar species: Peninsula cooter, Florida red-bellied cooter, Florida chicken turtle.

General range of species: The slider is a southern species of aquatic turtles that occurs naturally from New Mexico, Oklahoma, and Kansas to the East Coast of the U.S. It ranges as far north as Michigan, and some populations there are considered natural. There are many described subspecies. In Florida, the slider's natural southern range terminates around the Santa Fe River in North Florida. *Trachemys scripta scripta* ranges from northern Florida to southeastern Virginia and as far west as eastern Mississippi. It is not native to Florida but is now found throughout the state as a result of escaped or released pets. Both types of sliders are exotic to South Florida and Sanibel Island.

Island distribution: Both *T. s. scripta* and *T. s. elegans* are found on Sanibel. There are no records of either form on Captiva because of the lack of suitable freshwater bodies. *Trachemys s. scripta* x *T. s. elegans* (intergrades) are found throughout the island because of the presence of both subspecies.

Trachemys scripta elegans (Wied–Neuwied, 1838)
Red-eared slider (Actively introduced)[21]

This adult red-eared slider has left the water and dug a cavity into which she is depositing a clutch of eggs. All turtles lay eggs, and aquatic species of all kinds must leave the water to nest on land. *Bill Love*

Preferred habitat type(s): The slider is found in all freshwater and even in some brackish habitats. Its preferred habitat is lentic (still) water bodies with dense aquatic vegetation and soft muddy bottoms. *Trachemys* inhabits rivers, creeks, and spring runs, but its greatest densities are found in still waters.

Size: The adult male slider ranges from 9.4 to 24 cm (3.7 to 9.4 in) in carapace length; the female ranges from 19 to 30 cm (7.4 to 11.8 in). The maximum recorded SCL for *T. s. scripta* (the female is the larger of the sexes in many turtles) in the U.S. is 28 cm (11 in); for *T. s elegans* it is 28.9 cm (11.4 in). The average SCL of *T. s. scripta* at hatching is 32.0 mm (1.26 in), and that of *T. s. elegans* is 30.0 mm (1.18 in).

Color: In *Trachemys s. scripta*, the carapace is green to black with yellow streaks. The plastron is yellow with dark blotches, although some can have little or no blotching. The skin is brown to black with yellow stripes. There is a yellow blotch behind each eye. In *T. s. elegans*, the skin and shell color is very similar to *T. s. scripta*, though, *T. s. elegans* tends to have a greener shell. The blotch behind the eye in *T. s. elegans* is red or orange rather than yellow. Intergrades between the two subspecies have a blending of the two colors.

Other characteristics: The *Trachemys s. scripta* has vertical yellow stripes behind the thigh as in the chicken turtle, *Deirochelys reticularia*. Hatchling and juvenile *T. s. elegans* have a bright green carapace. The older *T. scripta* tends to become melanistic (black), especially the male, and lacks any distinguishing color characteristics.

Diet: The slider is omnivorous. It eats most aquatic invertebrates, as well as fish, amphibians, and carrion encountered throughout the day. Aquatic plants are eaten both purposefully and accidentally while hunting for prey.

Reproduction: The slider produces one to five clutches of eggs per year, ranging from five to 20 eggs per clutch. The yellow-bellied slider produces leathery, elliptical eggs that measure 22.6 by 37.7 mm (0.89 by 1.48 in), while the red-eared slider produces pliable, elliptical eggs averaging 21.6 by 36.2 mm (0.85 by 1.42 in).

A hatchling yellow-bellied slider. Although its carapace may bear a resemblance to that of a young cooter, red-eared slider, and chicken turtle, the large patch of yellow behind the eye is characteristic of this species.
Daniel Parker

Life history: The slider is the most common basking turtle in much of the U.S. and is usually the most common turtle in a habitat when present. It is often seen crossing a road as it moves to a new water body. The slider is highly adaptable and a survivor. The red-eared slider has been introduced all over the world as a result of the pet trade. There are now populations on all continents except Antarctica.

Population status: The yellow-bellied slider is very common on Sanibel Island. This turtle is found throughout the freshwater wetlands and in some brackish waters. The red-eared slider is less common on the island; however, this turtle's numbers have been increasing in recent years. Its reproductive interaction with the yellow-bellied slider is clearly recognized because intergrade individuals (*T. s. scripta* x *T. s. elegans*) are regularly found on Sanibel. The absence of a native, midsize, omnivorous freshwater turtle in South Florida has made colonization by this species relatively easy.

The carapace of a hatchling "dime-store turtle" is finely pinstriped with hieroglyphic-like lines. This species was once produced commercially in countless numbers for the pet trade until the fear arose that *Salmonella* infections were being transferred to young pet turtle owners by their turtles. This resulted in strict regulations enacted in 1975 prohibiting the commercial sale of turtles less than 10.2 cm (4 in) in carapace length. *Tim Walsh*

Threats: Fish, reptiles, birds, and mammals prey on the slider, from eggs to adults. Hatchlings and adults are collected for the pet trade.

Comments: In 1963, two University of Florida students visited Sanibel Island during a lengthy herpetological collecting trip in Southwest Florida. Somewhere in Florida, well to the north of Lee County, they collected a series of 16 mixed-sex adult yellow-bellied sliders. By the time they arrived on Sanibel Island, these turtles were in poor shape, and the students decided to release them immediately in a mosquito control ditch behind what then was the Villa Capri Motel on Periwinkle Way. The late owner/manager Marshall Tabbachi gave his permission, and the yellow-bellied slider was thus introduced to Sanibel Island. This waterway is connected to the Sanibel River, and the rest is history.

Range Map 09. Release point of *Trachemys s. scripta* on Sanibel Island, 1963.

In June 2009, a *Trachemys s. scripta* was photographed in the act of nesting in a hard-packed shell driveway on Periwinkle Way (R. Averill, pers. comm.). This property abuts an old spoil pond behind Periwinkle Way City Park. Later, a group of eight 2-day-old hatchlings that had emerged from this nest were examined by the senior author. The hatchlings were devoid of any external characteristics that would suggest they were intergrades between *T. s. scripta* and *T. s. elegans*.

A subadult slider from Sanibel Island that is an intergrade between *Trachemys s. scripta* and *T. s. elegans*. Note the red tint of the yellow lateral head patch behind the turtle's eye. This coloration represents the "red ear" characteristically associated with *T. s. scripta*. Intergrade specimens of *T. s. scripta* and *T. s. elegans* have become quite common in many areas on Sanibel Island. An *intergrade* is the offspring of parents of different subspecies, as opposed to a *hybrid*, which is the offspring of parents of different species. Chris Lechowicz

The red-eared slider, the famous "dime store turtle," came to Sanibel Island as a result of the flourishing turtle pet industry, and specimens were showing up on the island by 1975. The species is now very common and found throughout the freshwater wetlands of the island.

~ORDER CROCODILIA—THE CROCODILIANS~

Alligator mississippiensis (Daudin, 1802 "1801[22]")
American alligator

A large adult male alligator photographed from the Wildlife Drive of JNDDNWR while it was basking in the shallows of the refuge's West Impoundment on a sunny winter day. *Jim Fowler, Sanibel-Captiva Nature Calendar*

Other common names: Alligator, gator.

Similar species: The native American crocodile (*Crocodylus acutus*) is similar in general appearance and behavior to the American alligator (*Alligator mississippiensis*), as is the much smaller introduced spectacled caiman[23] (*Caiman crocodilis*). The latter is established in Southeast Florida. There is only one other species of true alligator, the smaller Chinese alligator (*Alligator sinensis*), which is restricted to China and considered critically endangered.

General range of species: The American alligator's range is confined to the southeastern U.S. Populations occur from the eastern halves of North and South Carolina southward throughout all of Florida. It occurs westward through the coastal states to eastern Texas and north into the extreme southeastern corner of Oklahoma and the southwestern corner of Arkansas. Florida and Louisiana contain the largest concentrations; recent estimates suggest that as many as 2 million alligators may now reside in Florida alone.

Island distribution: The alligator occurs on both islands, although availability of the reptile's preferred habitat restricts the majority of the permanent resident population to Sanibel Island. It is found throughout the interior wetland system and some aquatic habitats in the mangrove forest of JNDDNWR. Alligators regularly enter the mangroves and some seasonally leave the freshwater interior temporarily and forage in the surf along the Gulf of Mexico beach.

Preferred habitat type(s): The preferred island habitat of the alligator is the Sanibel Slough system and subdivision and golf course ponds. Many of these are interconnected and provide sufficient territorial space. In the interior wetland system of Sanibel, the alligator's individual territories are best measured linearly because of the former mosquito-control ditching program, as opposed to consolidated open water or marsh acreage.

Size: Hatchling alligators average 23 cm (9 in) on Sanibel Island. Under optimal conditions in the wild, they grow rapidly; it is possible some may reach sexual maturity on the islands at eight years of age. Reptiles are thought to continue to grow throughout their lifetimes, but the growth rate slows considerably after the average size is reached[24] In alligators this would be 3.6 m (12 ft) for males and 2.4 m (8 ft) for females. The largest American alligator ever recorded was killed in 1890 in Louisiana and measured 5.8 m (19 ft 2 in) (McIlhenny 1934), though this measurement is considered invalid by most herpetologists.

According to the FFWCC, the record length in Florida is a 4.37-m (14-ft 3.5-in) male alligator killed on Lake Washington along the St. Johns River in Brevard County during a state-sanctioned alligator hunt in 2010. It is unclear if the FFWCC is touting this specimen as the largest of the species ever documented in Florida, or just the largest ever measured during the sponsored alligator hunts. There is, however, another Florida record, for which documentary evidence is elusive, of a 5.3-m (17-ft 5-in) alligator killed in Lake Apopka in 1956 (Neil 1971). This specimen's length was taken shortly after its demise and apparently later supporting evidence of length was extrapolated from its skull size. It appears that the FFWCC does not accept this record.

According to the late George Campbell, the largest American alligator ever measured on Sanibel Island was 5.2 m (17 ft) (Campbell and Winterbotham 1985). Campbell names two reputable biologists who verified the measurement of this alligator that some island residents had fondly named Bismarck after this giant was shot and killed in 1973. Unfortunately, Bismarck's length and supporting morphological measurements were never published in the scientific literature, and today this animal's size is discounted by many herpetologists.

Little is known about the longevity of wild alligators. Some in captivity have exceeded 50 years of age.

Color: Hatchling alligators are brightly marked with bands of black and yellow. These vivid hues are lost with growth, and in time the adult color becomes a drab olive-gray.

A hatchling American alligator standing upright and defensively poised. The contrasting black and yellow markings provide excellent camouflage. These markings disappear as the alligator ages, but even 11-foot alligators may sometimes reveal a faint trace of the juvenile pattern. *Tim Walsh*

Other characteristics: The alligator is a highly vocal animal. Even as preemergent hatchlings, they have the ability to produce a moderately high-pitched chirplike sound, known as a grunt. When coming from inside an egg in the nest, these sounds may be muted but are still audible to the attending female. This vocalization is common among post-hatching siblings and is used for communication with their mother. This sound is also used as an alarm, a noisy distress cry when the alligator is threatened. In many instances, the female provides an excellent level of protection over her developing eggs or successful brood, although females do not universally behave defensively to protect their nests.

The sound-producing capability changes over time and becomes much lower in tone with age. Individual adult alligators even vocalize among themselves using this type of communication. Both sexes are capable of loudly bellowing, but it is most often the male, or bull, that uses this unique vocal capability, often as part of his territorial advertising or while courting a female. When a female alligator is defending her nest or young, she will sometimes advance toward a human intruder with jaws open while at the same time emitting a loud threatening hiss.

Diet: The alligator is an opportunistic ambush predator; it is considered a primary carnivore and is at the top of the wildlife food chain in Florida's non-marine ecosystems. Hatchlings progress from a diet of insects and spiders to small fishes, amphibians, and reptiles. By the time they are adults, alligators consume turtles, snakes, nestling birds that fall into the water, small mammals such as raccoons, and larger prey such as unwitting deer and young cattle. Small alligators themselves fall prey to larger alligators. Alligators are also known to ingest small stones and heavy wood knots. Once swallowed, these objects remain in the animal's stomach and are known as gastroliths. They are believed to have two functions: 1) to provide food-grinding assistance, similar to the reason birds swallow grit; and 2) to help regulate their buoyancy in the water, as modeled by Henderson (2003). Humans are not considered natural prey of alligators. In many cases where alligators have attacked people, a dismembered body part may have been eaten.

Reproduction: Alligators go through an interesting courtship during which males and females display a variety of social interactions. During these mating rituals, they copulate multiple times. Fertilization is internal. The eggs of the alligator are hard-shelled and elongated. An average egg measures 42.5 by 73.7 mm (1.67 by 2.9 in) and weighs 79.4 g (2.8 oz). In Florida, nest building and egg laying usually occur between late March and May. Sanibel Island's alligators fit into this general time frame, though hatchlings that had just pipped the eggshell and were still showing prominent umbilical scars were observed as early as February by George Campbell (Campbell and Winterbotham 1985).

Life history: Hatchling alligators initially remain close to their mother for the protection she provides for up to two years. By the time they are approximately 61 cm (2 ft) in length, the unity of the pod crumbles and the young alligators disperse. By the time maturity is attained, both sexes have established their individual territories to which they restrict their movements unless environmental parameters change and they must relocate. In Florida, female alligators dwell in about a 50-hectare (ha) territory, while adult males range through territories that can reach 480 ha in size. Their territories are not that expansive on Sanibel Island.

The adult alligator excavates burrows, or dens (or caves, as they have been traditionally called on Sanibel Island). Caves are usually located in the bank of the body of water in which the alligator dwells. In the first half of the 20th century, the area that is delineated by the yellow dashed line and identified as C on Map 7 was called "Gator Heaven" by some early island residents (H. Rhodes. pers. comm.). Men frequently hunted alligators on Sanibel for the illicit hide trade and knew about the many large alligators in Gator Heaven, but it was mostly an inaccessible wetland system and difficult to hunt there.

George Weymouth and LeBuff examined enormous cave openings in West Government Pond in Gator Heaven, now within JNDDNWR. Some of these were positioned beneath red mangrove trees, which probably helped support the den's ceiling. Several openings around the margins of this pond measured more than 1.5 m (5 ft) across when examined when water levels were at a minimal level in 1964. The caves themselves were still flooded and the water level was just a few centimeters below the rim of the entrance. When a 3.6-m (12-ft) Calcutta bamboo pole was inserted into any one of several dens, its tip did not contact the end of the tunnel, nor encounter a resident. Elsewhere in Florida, some alligator (and crocodile) caves have been recorded that were 12 m (40 ft) in length. At the end of such a cave, the resident alligator forms a wide turning room, and there is typically an air chamber near the ceiling of this part of the cave. Crocodilians seek refuge in these caves when threatened or because of drought or extremes in temperature.

Population status: From all indications, the public visibility of large alligators is diminishing on Sanibel Island. This can probably be correlated with the existing COS alligator harvest policy. Recent trends in long-term, ongoing index-study alligator counts indicate a reduction in numbers. The American alligator is currently in trouble on Sanibel Island.

Alligator surveys are conducted by the JNDDNWR and SCCF twice a year. These surveys are carried out only on conservation lands and the golf courses on Sanibel Island. Private lakes and other private water bodies are not surveyed. Eye shines are recorded during nocturnal surveys, although length estimates are not recorded. Length estimates are recorded during diurnal surveys, which are mostly conducted on the three golf courses on the island.

Threats: The American alligator population had plummeted nationwide by 1960. In states with alligator populations, conservation agencies pushed for promulgation of tougher state statutes, but even with increased efforts law enforcement was unable to turn the rising tide of illicit trade in alligator hides that was sweeping across the Southeast. Without a significant change America would likely have witnessed the extinction of this important and uniquely American species. Sanibel Island was not immune to illegal harvest of alligators (LeBuff 1998).

Finally after much political procrastination, the American alligator was listed by the U.S. as Endangered in 1967, and its status began to change. The alligator's comeback was dramatic. Its population in Florida has grown to the extent that some communities have serious problems with nuisance alligators. Presently, there are no serious threats to the statewide population—outside of Sanibel Island.

Comments: The American alligator was removed from the U.S. Endangered Species List in 1987 when the USFWS determined the population had recovered. Today, it remains federally protected as Threatened because of its similarity in appearance with protected crocodilians. The state of Florida considers the alligator to be a Species of Special Concern.

In 1956, Fred Stanbury, head of the FGFWFC Wildlife Management Division, granted LeBuff a letter/permit that authorized him to capture, tag, and release alligators in Collier County, Florida. From early that year until late 1958, 1,500 alligators were caught, tagged, and released, mostly in the Big Cypress Basin of Collier County. After he relocated to Sanibel Island, the senior author continued working with alligators and managed the island's nuisance alligator program from 1959 until 1971, when he redirected his career to sea turtles. During that period, an additional 500 alligators were tagged or otherwise marked by LeBuff on Sanibel Island.

Sanibel alligators are cosmopolitan and range through all habitats, and some regularly enter the Gulf of Mexico (LeBuff 1998). The mosquito-control program and the coincidental spoil ponds and borrow canals that resulted from real estate development enhanced this crocodilian's habitat and spurred a population increase. At first, when alligators were considered nuisances on private lands, they were caught and tagged or marked for future identification. Then they were relocated to remote sections of the wildlife refuge. Prior to LeBuff's involvement, concerned residents would sometimes shoot problem alligators while a few others were caught and taken to wildlife exhibits on the mainland.

After 1971, the management of nuisance alligators was turned over to a series of state-permitted individuals. After LeBuff, the first volunteer administrator was George Weymouth. In 1973, after an agreement supporting a regional coalition to foster alligator protection was reached among island residents, the Southwest Florida Regional Alligator Association (SWFRAA) was organized, and George Campbell of Sanibel was selected as its chairman. Mark Westall later became the primary alligator coordinator for SWFRAA. It was near the end of his term, with alligator complaints increasing, that the nuisance alligator program was transferred to the COS, and the Sanibel Police Department was delegated the responsibility for nuisance alligator management. After its transfer from the private sector, the program continued to work relatively well—until 2001. That year, Sanibel Island vaulted into infamy because of its resident American alligators, and the community's love-hate relationship with alligators became unbalanced.

From 2000 to 2009 Sanibel Island was the site of four serious alligator attacks on humans, two of which resulted in fatalities. Sobczak (2006) provides accounts of some of these tragic incidents. Dogs that two of the victims were walking are believed to have been the target of the attacks, although from all evidence the most recent fatal incident was completely unprovoked. Immediately thereafter, with the insistence of the FFWCC, the Sanibel city council struck down the long-established alligator management policy and adopted that of the state of Florida. Henceforth, any alligator exceeding 1.2 m (4 ft) in length that appears to be a potential threat—that is, one that displays a set of inappropriate behaviors—is caught and killed by the state-licensed alligator trapper who serves Lee County. This became a very difficult time for alligators on Sanibel Island. The COS had to respond to public concern generated by the attacks and fatalities, and the result was a culling of any alligators larger than the dimensions given above. Unfortunately, many people continue to call the COS whenever they see an alligator in the water within their subdivisions, and the trappers come out to the island and remove it. Many of the animals do not qualify for removal and become victims of human hysteria.

One of the major issues with the new COS alligator policy is that the trappers are removing many of the large adult alligators that make up the island's breeding population. The JNDDNWR and SCCF refuse to allow trappers to take non-nuisance alligators from their lands. Most of the wetlands on the preserved tracts[25] are not very deep, and adult alligators prefer the deeper water found in manmade private water bodies. After the large alligators were removed from private ponds, the alligators on conservation lands were able to spread out and claim the recently vacated territory. Hence, over time, those alligators will also become candidates for removal by the trappers.

The scientific alligator capture/recapture tagging program that was initiated in 1959 was curtailed after 1973. Relocated alligators were later customarily marked by dorsal caudal scute removal to identify them in a codified system. Early on, tagging indicated that for the most part, Sanibel alligators remained on the island. When relocated to other parts of the island, adults usually returned to the points of capture in their established territories within days. The longest straight-line movement for a Sanibel alligator was 16.7 km (10.4 mi) when a tagged individual moved of its own accord from the center of Sanibel Island to the southern section of Estero Island[26] in 1966. Campbell (1985) provided some additional data on the movement and growth rates within the Sanibel Island alligator population.

A significant contribution to the conservation of the American alligator on Sanibel Island, and ultimately in Florida, was made in 1974. The reptiles were losing their fear of humans and would boldly approach anyone near the water in anticipation of being tossed a morsel. Some employees at island grocery stores thought themselves to be experts in this reptile's dietary needs and were recommending packaged alligator cuisine to anyone who asked. It was remarkable just how quickly visitors learned how best to take an alligator's portrait: attract it with a morsel of floating food and get it to move in close. Marshmallow sales at grocery stores skyrocketed during this period.

LeBuff long recognized the potentially dangerous situations that island residents and visitors caused when they actively attracted and fed alligators. Then a member of the Sanibel City Council, he drafted an ordinance prohibiting feeding American alligators within the incorporated limits of the city, and the council unanimously adopted Ordinance 75-29. This became an effective tool in local alligator management. The FGFWFC saw the merit of the Sanibel policy and picked up on the concept in 2006, adopting Rule Number 68A-25-001. With this rule added to the Florida Wildlife Code, feeding wild alligators was banned in Florida. Enforcement is another issue.

Crocodylus acutus Cuvier, 1807
American crocodile

An adult American crocodile photographed as it leisurely makes headway at the water's surface. This is the typical crocodilian way of swimming while on or beneath the surface; the crocodile's appendages are relaxed and held close to the body as the powerful tail propels the animal through the water. *Bill Love*

Other common names: Florida crocodile, Everglades crocodile[27], croc.

Similar species: Similar in general appearance to the alligator and the introduced spectacled caiman (*Caiman crocodilis*), which has become established in Southeast Florida.

General range of species: In Florida, the American crocodile ranges along both coasts, including estuaries and barrier islands, and throughout the Florida Keys. The northern extremes of its known range are near Tampa Bay on the Gulf Coast and Lake Worth to the east.

South of Florida, the American crocodile is distributed through most of the major Caribbean islands and both coasts of central Mexico, then southward along the coasts of Central America to northern South America. On the South American continent, this species occurs in easternmost Venezuela on the Caribbean seaboard and northern Peru along the shore of the Pacific Ocean.

Island distribution: The only specimen known to reside long-term on Sanibel Island (1979-2010) preferred the western end of JNDDNWR. Infrequently, this crocodile safely negotiated crossing Sanibel-Captiva Road and ventured into the island's interior freshwater wetlands.

Preferred habitat type(s): In southern Florida, the habitat of the crocodile is usually associated with the coastal distribution of the red mangrove (*Rhizophora mangle*). Most Gulf Coast American crocodiles recorded near the islands were captured north of the CHES or at its northern periphery. The majority of them, beginning in the 1940s and lasting at least until 1955, were collected in the vicinity of Lemon Bay in Charlotte County and relocated to the collection of a privately owned wildlife attraction in Bonita Springs (LeBuff 2010). As of 2013, a nonbreeding group of adult American crocodiles resides near Marco Island in Collier County.

This crocodilian typically uses habitat associated with the estuary of Florida Bay and the mangrove forests associated with Cape Sable in Everglades National Park. The preferred reduced salinity levels of the waters of this estuary are properly maintained because the system receives freshwater originating inland in the Everglades. Manmade freshwater pools adjacent to the cooling canals of the Turkey Point Nuclear Power Plant in Dade County are regularly used by adult females as short-term nursery ponds for their hatchlings.

Size: This is one of the world's largest crocodilians. Hatchlings average 23 cm (9 in), and from that diminutive size the longest adult ever measured reached 7 m (23 ft) in total length. This huge specimen was recorded in South America. In Florida, the record length for a male American crocodile is 4.6 m (15 ft). The deceased female Sanibel Island crocodile was one of the largest of her sex ever measured anywhere—3.66 m (12 ft) in total length[28]. This specimen's pertinent measurements were collected postmortem by Lechowicz.

Color: At hatching, American crocodiles are two-toned: a predominantly light tan-olive green ground color with narrow black bands or blotches on the dorsal and ventral sections of the body and tail. The head has a light base color and is flecked with black or dark brown. Most of the body pattern disappears with growth. The color of the adult American crocodile is usually described as a tan-hued gray when the animal is dry and a similar but darker tone when wet.

Other characteristics: The teeth of the American crocodile are much different in appearance from those of the American alligator. Many of the 70 teeth are exposed and visible when the crocodile's mouth is closed. The fourth tooth of the mandible (lower jaw) is totally visible when the mouth is shut tight, and the upper jaw is grooved to provide clearance for this tooth. It is not uncommon in old males for the front teeth of the lower jaw to protrude several millimeters through the upper jaw when the mouth is closed. The American crocodile regularly basks on land with its mouths agape. Crocodilian specialists consider this behavior to be a method of thermoregulation that assists the crocodile's digestive processes. The open mouth may also help to reduce growth of unwanted organisms in the mouth.

Contrary to folklore, the lower jaws of both alligators and crocodiles are hinged alike; the bottom jaw is the moveable one on both animals.

Diet: As is the American alligator, the American crocodile is a carnivorous apex predator. Young crocodiles consume a variety of small organisms: spiders, insects, aquatic invertebrates such as juvenile crabs, and vertebrates such as minnows and amphibians. The adult American crocodile takes a variety of prey: these include crabs, fish, and, when the opportunity arises, mammals and birds. There is no record of an American crocodile attacking a human in Florida, although there are documented cases of attacks in Central and South America.

Reproduction: Mating occurs in the water, and the female will later usually build a mound nest close to where she had nested previously. The mound is usually composed of soil, but sometimes vegetation may be included in its construction. In Florida, the female American crocodile may alter her nest-construction behavior. When the soil of a nesting site is elevated and well drained, she may simply excavate a hole in the ground, as a turtle does, into which she drops her eggs. By carefully using her rear legs, she scoops out an egg chamber in the center of the mound or the sandy area and deposits an average of 38 eggs. These are elongated and have hard, calcified shells that average 44.0 by 70.0 mm (1.73 by 2.75 in) and weigh 86 g (3.0 oz).

A hatchling American crocodile at the Turkey Point Nuclear Power Plant in Dade County. There are no records of viable natural crocodile nests along the Florida Gulf Coast north of the western Cape Sable region in Everglades National Park. *Bill Love*

Life history: Following an 85-day incubation period, the female crocodile frequently visits her nest and listens for telltale grunting sounds emanating from inside the buried eggs. This sound is generated by preemergent hatchlings just before they are ready to pip their eggshell and escape the confines of this imprisoning capsule. At this stage, their survival is critical, so the female excavates the nest to release the full-term hatchlings; she even takes partially pipped eggs and hatchlings in her mouth and carries them to the water. This maternal behavior may also permit the young to be released in the event the leathery inner layer of the egg has hardened. Eggshell hardening can occur if incubation has taken place during periods of exceptionally dry weather. After being transported to the water, a pod of hatchling crocodiles may be moved a considerable distance from the nest site by the female. She will pick them up and carry them in her mouth or lead them to better nursery habitat where the lower salinity of the water is more conducive to their development; this enhances their survival potential. Once her young are secure in such a "nursery" habitat, the female American crocodile forgoes any additional protection of her offspring and abandons them. Crocodiles thus lack a predisposition for parental care, which their relative, the alligator, has developed to a very high level.

Population status: The American crocodile population in Florida reached its historical peak in the first quarter of the 20th century after completion of the Overseas Railroad to Key West. Railroad-grade fill from borrow canals increased the availability of nesting sites, and the growth of the Florida Bay crocodile population soared. By the late 1930s, this population had declined considerably because of hunting pressures, commercial collections for zoological exhibits and museums, egg collection to provide hatchlings for the pet trade[29], and road-kill mortality. Finally, after 1947, with the establishment of Everglades National Park and Florida prohibiting the take of American crocodiles in Monroe County, the population began to rise again. Years later, after the American crocodile was protected under the ESA, and when new habitat in Dade County became available during the 1970s, Florida's population of the American crocodile expanded once more. Slowly, this reptile dispersed from the center of its population in upper Florida Bay. Over time, many adults moved up the Gulf Coast and established new territories at places such as Marco Island, Bonita Springs, and Lemon Bay near Englewood.

Threats: Since addition to the U.S. Endangered Species List in 1975, the survival outlook of this crocodile has brightened. As previously mentioned, its population has surged, and its range, at least along the Gulf Coast, is expanding. Despite global climate change, this crocodile likely will be restricted to extreme southern Florida well into the next millennium because it is seriously threatened by cold winter temperatures. The 2010 cold front affected many specimens in Florida Bay and the media reported about 100 crocodile fatalities because of cold stress in that region.

Comments: The status of the American crocodile was reduced from Endangered to Threatened in 2007 under guidelines of the U.S. ESA. Florida continues to classify the American crocodile as Threatened.

The distribution of this crocodile along Florida's Gulf Coast has been the subject of earlier discussions (LeBuff 1957). Since then, there has been a gradual increase in the region's crocodile population. In recent years, individual American crocodiles have been discovered as far north as Tampa Bay, Manatee County (Klinkenberg 2008), and more recently in Pinellas County. Over the past several decades, small groups of this crocodilian have been discovered in Lee and Collier counties south of Sanibel Island. In the past, with records going back into the early 1950s, adult crocodiles have been infrequently caught and relocated from the vicinity of Lemon Bay in Charlotte County and Big Sarasota Bay in Sarasota County. The most recent of these as of 2012 was a 2.4-m (8-ft) female captured near Englewood, Charlotte/Sarasota counties, and released in JNDDNWR. After being released, the animal adjusted and moved eastward a few kilometers. In 2013 it continues to reside in the Dunes Country Club, Sanibel Island.

In 1936, an adult American crocodile was caught in a commercial fisherman's gill net in McIntyre Creek, a tidal mangrove creek now within JNDDNWR (J. Lamb, Jr., pers. comm.). The presence of another American crocodile would not be documented on Sanibel until 1979, when an amateur photographer captured the image of a basking adult near the western boundary of the refuge (Campbell and Winterbotham 1985). On 6 June 1980, naturalist George Weymouth sighted a crocodile on JNDDNWR. He notified the refuge office, and LeBuff responded to the report. This site, just inside the JNDDNWR boundary, was extremely close to where the 1979 American crocodile had been photographed. At the time, the animal was extremely timid, but LeBuff managed to capture 14 photographs—all of them just of the top of the crocodile's head.

Over time, this large specimen became accustomed to the close approach of people. In 1997, an amazing thing happened. This fine specimen turned out to be a female and began to construct nest mounds and lay eggs in the Sanibel Bayous Subdivision, a site quite close to the location where it had been originally photographed. Through the succeeding decades, this crocodile periodically nested but deposited infertile eggs. The most recent clutch deposited by this crocodile, in 2009, was abandoned by the female, (J. Combs, pers. comm.).

The American crocodile is not known to nest successfully on the Gulf Coast north of Cape Sable. However, a large captive group in Bonita Springs, Lee County, collected in the Florida Bay region between 1937 and 1947, produced viable eggs until at least 1954 (LeBuff 2010).

The Sanibel Island crocodile was observed in rather intimate association with large male alligators on at least two occasions. The first was when LCMCD employee Roger Zocki (pers. comm.) observed the crocodilians within 7.5 m (25 ft) of each other in a borrow canal along the Lee County Electric Cooperative's right of way, which passes through JNDDNWR. When it was originally excavated, this canal crossed the westernmost pool that is seasonally connected by water to West Government Pond (See Map 7, page 66). The next time was more remarkable, when a volunteer at the refuge had the unique opportunity to snap an outstanding photograph of the resident crocodile while she engaged in an apparent nonviolent territorial confrontation with a large American alligator (T. Baxter, J. Combs, pers. comm.).

Through the years, her eggs were consistently infertile. This raises an interesting question: did the two individuals ever attempt copulation?

Frank Mazzotti (pers. comm.) has observed alligators and crocodiles court in captivity and interact socially in the wild. He blames the Sanibel Island crocodile's lack of a conspecific mate (one of its own species) as responsible for the infertility of her eggs; it also has been suggested that sperm incompatibility may be the reason (LeBuff 2007). LeBuff last personally observed this crocodile on 28 September 2009, while in the company of JNDDNWR manager

Paul Tritaik. The specimen had somehow entered a fenced yard and was temporarily trapped. She later maneuvered and escaped through a gate that was left open for that purpose.

This photograph was taken from the JNDDNWR Wildlife Drive on an early morning in February 2005. The crocodilians were in a borrow ditch, part of the refuge's east impoundment—at a location about 0.5 km (0.3 mi) north of Sanibel-Captiva Road. The crocodile (left) was seen first when the alligator (right) approached it. Both animals swam with their heads just above the water surface. As they came closer to one another the crocodile raised its head and showed its teeth, whereupon the alligator reacted quickly by opening its mouth. Immediately, the crocodile's head was on top of the alligator's, as captured in the photograph, and the alligator turned and swam quickly away in the direction from which it came. The encounter was over in less than a minute. There was no submerging; no attempt to bite each other; no wild splashing in the water—the alligator just hurriedly swam away, retreating back toward the impoundment. The crocodile followed for a few meters, stopped, and then watched the alligator leave the area. Both were about the same size, so the crocodile did not have any size advantage. Both the alligator and crocodile soon disappeared among the roots of the mangroves. The photographer was very pleased to have witnessed this behavior and record this unique encounter. (T. Baldwin, pers. comm.)
Theresa Baldwin

This fine specimen was discovered dead on the bank of the Sanibel River on SCCF land by the staff member Dee Serage-Century on 26 January 2010. Lechowicz collected postmortem measurements of this crocodile at that time. The crocodile's demise was attributed to the extremely cold weather at the time, although age may have been a key factor. Its skeletal remains were rearticulated and are on display at the JNDDNWR Education Center.

Another American crocodile was sighted north of Sanibel Island in September 2011. This crocodile was photographically documented basking on the shoreline of Old Tampa Bay in a subdivision in St. Petersburg, Pinellas County. Because of the unfounded concerns of nearby residents, this crocodile was a candidate for capture and relocation to a suitable habitat somewhere farther south within the historic range. If this occurred it was apparently not publicaly disseminated by FFWCC officials. The most recent American crocodile to be verified north of Tampa Bay was an 11-footer that was captured by a state-licensed trapper in July 2013. This specimen was collected in the extreme northern end of Lake Tarpon, also in Pinellas County, 35.5 km (22.06 mi) north of where the smaller 2011 specimen was taken. According to the published account, it was temporarily relocated to a rehabilitation facility before an expected release in South Florida. The latter individual is now the northernmost record for this species on the Florida gulf coast.

~ORDER SQUAMATA—THE LIZARDS~

Agama agama africana (Hallowell, 1844)
West African rainbow lizard (Passively introduced)

A male West African rainbow lizard. As the population of this exotic lizard expands on the nearby mainland, more individual agamas will reach the barrier islands by vehicles and likely prosper. *Daniel Parker*

Other common names: African rainbow lizard, red-headed agama, common agama.

Similar species: None on Sanibel or Captiva.

General range of species: *Agama agama africana* is found naturally in sub-Saharan Africa.

Island distribution: As of 2013, this species is known from one specimen from the far western end of Sanibel Island. The closest population to Sanibel Island is in Punta Gorda, Charlotte County, Florida.

Preferred habitat type(s): *Agama agama africana* is found in the bush in Africa. It is not typically found in barren sandy areas, but in places with ample vegetation. In Florida, it characteristically lives on the sides of buildings or on rock piles. As in Africa, it quickly retreats to a tight space under rocks or any other available cover.

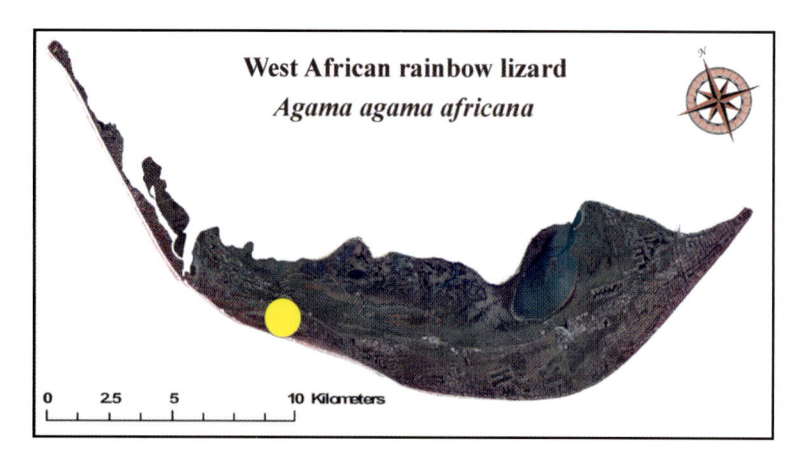

Range Map 10. The first West African rainbow lizard on Sanibel or Captiva was documented at this location on Sanibel Island in 2006.

Size: A medium-sized lizard, 12.5 to 30 cm (adults are usually 5 to 12 in long). This lizard's maximum recorded length is 35.5 cm (14 in)[30].

Color: Adults are sexually dimorphic in color. The male is blue with a red head, whereas the female is mostly brown to gray in warm temperatures. When the temperature is cooler, both sexes are similarly colored—that is, mostly brown or gray.

Other characteristics: The West African rainbow lizard has keeled scales and strongly developed legs. Its tail is very long and stiff, and the tail of the male *A. a. africana* has an orange segment in the middle and a blue tip.

Diet: *Agama agama africana* eats mostly insects, but will sometimes eat grasses and berries.

Reproduction: The male of this species often has several mates. It bobs its head like other lizard species to attract females and to discourage rival males from intruding into its territory. The female deposits multiple clutches of eggs with as many as 12 eggs per clutch.

Life history: This agama is diurnal and can withstand very high temperatures. Dominant males set up territories and will fight with other males that enter their territories. They head bob to show dominance. This lizard is found mostly near the ground, although it is an excellent climber. As with all agamas, it is very skittish when approached.

Population status: The population of this agama on Sanibel Island appears to be extirpated. The one documented individual likely died that same year.

Threats: *Agama agama africana* is not native to Florida and does not appear to be reproducing on either island.

Comments: The East African rainbow lizard has been introduced to at least five counties in Florida (Enge et al. 2004) and was recorded on Sanibel Island in 2006 (Lechowicz[31] 2009).

A specimen was first observed and photographed on the island by Malcolm Harpham on 5 May 2006.

This is another favorite exotic lizard common to the pet trade, but its arrival on the island was likely passive, coming via landscape nursery plants. However, the first of these lizards to reach Sanibel Island could have been an escapee or even may have been liberated by its owner. As with the knight anole and the curly-tailed lizard, it is likely the introduction of the agama was facilitated by the massive amount of vegetation brought to Sanibel and Captiva islands to restore landscaping after Hurricanes Charley and Wilma in 2004-2005. All three species are common in the plant nurseries in Homestead and Miami. This explains why the three species have been observed in the same general area where most new landscaping was undertaken.

Anolis carolinensis (Voigt, 1832)
Green anole

The attractively colored native green anole was popular when the American public's interest in herpetology intensified in the 1940s. This lizard helped launch the careers of many herpetologists. *Bill Love*

Other common names: American chameleon, Carolina anole.

Similar species: Brown anole.

General range of species: *Anolis carolinensis* is found in the southeastern U.S. from central Texas to Florida to Tennessee and North Carolina.

Island distribution: Found throughout Sanibel and Captiva islands in natural environments and sometimes manmade habitats.

Preferred habitat type(s): The green anole lives primarily in vegetation high above the ground.

Size: A relatively small lizard, 12.5 to 20 cm (adults are usually 5 to 8 in long), with a maximum recorded length of 20.3 cm (8 in).

Color: This anole is primarily green and has the ability to change its dorsal color to brown. The dewlap of the male is pinkish red.

Other characteristics: *Anolis carolinensis* has a longer head than *A. sagrei*. When in its brown phase, the green anole may have visible markings on its back.

Diet: This lizard eats insects, various invertebrates, amphibians, and reptiles.

Reproduction: Throughout the summer months, the green anole deposits multiple egg clutches, each containing one or two eggs. These are deposited at two-week intervals until 10 eggs, on average, have been produced.

Life history: The green anole is a diurnal lizard that enjoys basking in the sun on vegetation. The male sets up a territory and defends it by flapping his dewlap to warn other males of his occupation of the territory. A male will also bob its head and body up and down, resembling push-ups, for the same reason. The green anole was once a commonly seen lizard in South Florida. In the past, it was the common lizard that residents saw on their pool enclosures, landscaping, and homes, but it now has been typically replaced by the brown anole.

Population status: The population of this native anole on Sanibel and Captiva islands appears to be in decline. This is no longer a common species on the islands—far from it. It is locally common in a few areas, but is most often seen as a solitary individual among the numerous brown anoles.

Threats: The main threat to the green anole is the invasive brown anole. The latter prefers to be about two meters off the ground and has been very successful at colonizing the lower vegetation heights previously frequented by the green anole. The brown anole outcompetes the native green anole in various ways. The latter is much more aggressive and regularly produces more offspring per year than the native anole.

Comments: *Anolis carolinensis* was once commonly known as the American chameleon. This delicately green-colored lizard was sold at circuses and carnivals to thousands of children. It was typically fastened with string to safety pins to be worn on children's clothing. The green anole commonly occurred throughout Sanibel Island in 1959. Today, its abundance appears to be reduced and individuals are observed infrequently. It is not uncommon to observe confrontations, even physical fights, between brown and green anole males.

Anolis equestris equestris Merrem, 1820
Western knight anole (Passively introduced)

A young adult western knight anole. This is the largest of the anoles and is extremely territorial. This individual has its mouth wide open, a typical defensive posture. *Bill Love*

Other common names: Cuban knight anole, Cuban anole.

Similar species: The juvenile knight anole can appear similar to the adult green anole.

General range of species: *Anolis equestris* is found in Cuba but has been introduced into Dade and Broward counties and has quickly spread around southern Florida.

Island distribution: As of 2013, the knight anole has been documented only on Captiva Island.

Preferred habitat type(s): The knight anole is an arboreal species that spends most of its time in shady trees and shrubs. It will bask on the ground but quickly run up a tree when disturbed.

Size: A very large anole, 33 to 51 cm (adults are usually 13 to 20 in long). The maximum recorded length for this lizard is 55 cm (22 in).

Color: The knight anole is primarily lime green with a yellow bar under each eye. Its color can darken to almost all brown or lighten with the changing ambient temperature. The male has a pale pink dewlap.

Other characteristics: *Anolis equestris* has enlarged toe pads on each toe to help it run up and down vertical surfaces. It has a strong, elongated jaw.

Diet: It eats mostly insects, but will also eat smaller amphibians, reptiles, and even baby birds.

Reproduction: *Anolis equestris* breeds in the summertime. The female can deposit as many as four clutches containing one to two eggs each.

Life history: The knight anole is the largest anole in the world. It is highly territorial and bobs its head and shows its dewlap like other anoles to chase away intruders.

Population status: As of 2013, we do not know whether the knight anole is reproducing on Captiva, but it is likely.

Range Map 11. The occurrence of the exotic knight anole on Sanibel or Captiva was first documented at this location on Captiva Island in 2009.

Threats: Since this is an exotic species, the threats to its existence in Florida have not been considered. It is a threat to the green anole because both are diurnal, and the knight anole is much larger than the native species; because of this, the green anole becomes prey.

Comments: As with the agama and the curly-tailed lizard, the introduction of this species to Captiva Island likely was caused by the massive amount of plants brought to the barrier islands after Hurricanes Charley and Wilma in 2004-2005. In June 2012, another knight anole was photographed by Lechowicz on Captiva Island.

Anolis sagrei Duméril and Bibron, 1837
Brown anole (Passively introduced)

The brown anole represents an early success story for a nonnative species that extended its range from the nearby West Indies to the U.S. This species is very abundant and serves as an important part of the prey base of native wildlife species, including the barrier-island herpetofauna. *Bill Love*

Other common names: Cuban anole, chameleon.

Similar species: Brown-colored green anoles.

General range of species: *Anolis sagrei* is native to Cuba and the Bahamas. It has been introduced in the Caribbean, Taiwan, Florida, Georgia, California, and Hawaii.

Island distribution: *Anolis sagrei* is found throughout Sanibel and Captiva islands in all habitat types.

Preferred habitat type(s): The habitat of the brown anole includes both tree trunks and the ground in disturbed and undisturbed sites. It is not particular about habitat except that it prefers lower heights in the tree canopy.

Size: A relatively small lizard, averaging 12.7 to 20.3 cm (adults are usually 5 to 8 in long), and attaining a record length of 21.3 cm (8.4 in).

Color: *Anolis sagrei* is capable of changing its body color and can be light brown to almost completely black. Most often it is some shade of brown with white undersides.

Other characteristics: Sometimes the brown anole has triangular patterns on its dorsal side or varying degrees of spotting or lines. The tail is laterally compressed, and its dewlap is yellow to red. The males often has a crest down its spine and is larger and heavier than the female.

Diet: The brown anole primarily eats insects, but will also eat other small invertebrates and juvenile lizards.

Reproduction: The male sets up a territory for breeding and aggressively defends it from other brown anoles. Dewlap displays are used to attract females. The brown anole breeds in the spring and summer in Florida. The female has been reported to deposit many clutches of one to two eggs periodically throughout the summer.

An adult male brown anole in an aggressive mode displays its brightly colored dewlap.
Daniel Parker

Life history: The brown anole is semi-arboreal, meaning it mostly stays low in trees and vegetation. The trunks of trees or walls are its preferred perching areas. It spends most of its day in an upside-down vertical position hunting insects and basking in the sun.

Population status: Its abundance on the islands appears to be increasing. It rapidly becomes the dominant lizard in all areas it colonizes. In some areas of great abundance, it may increase its range vertically by pushing the green anole higher up in the canopy.

Threats: Since this is an exotic species, the threats to its existence in Florida have not been considered.

Comments: The very abundant brown anole did not appear on Sanibel or Captiva islands until about 1972. The species was added to the refuge amphibian and reptile list that year.

Once this exotic was minimally established, the population exploded, and today this lizard occurs in all major ecosystems. Because the brown anole resides mostly at eye level or lower, humans tend to see it more than the green anole. The brown anole is extremely invasive and can become established very easily in warm climates. Its trek northward has not stopped, and it is not known how far north this species can be successful as temperatures become increasingly warmer during winter months because of climate change.

Aspidoscelis sexlineata (Linnaeus, 1766)
Eastern six-lined racerunner

An adult six-lined racerunner. These speedy lizards occupy open, sparsely vegetated, dry habitats, such as the foredune areas of the upper Gulf Beach Zone. *Daniel Parker*

Other common names: Racerunner, field streak.

Similar species: Southeastern five-lined skink.

General range of species: *Aspidoscelis sexlineata* ranges from northern Maryland and west to Missouri and East Texas, and through the contiguous southeastern states as far south as the Florida Keys.

Island distribution: The six-lined racerunner is common on Sanibel and Captiva islands in disturbed and undisturbed uplands.

Preferred habitat type(s): This lizard is an upland species. It prefers high and dry areas, including the dune area of the beach, grasslands, and woodland edges.

Size: Moderate size, 15 to 22.8 cm (adults are usually 6 to 9 in long). The record length for this species is 27 cm (10.6 in).

Color: The six-lined racerunner is gray, brown, or green on the dorsal side, and white on the ventral side. The ventral side of the male is blue-hued.

Other characteristics: *Aspidoscelis sexlineata* has six white to yellow stripes on the dorsal side.

Diet: This carnivorous lizard eats insects, spiders, and other invertebrates.

Reproduction: The six-lined racerunner breeds in the spring and deposits as many as five eggs in the summer.

Life history: *Aspidoscelis sexlineata* is diurnal and terrestrial, meaning it is active only during the day and spends its time on the ground. It is most active during the hottest part of the day. At night it seeks refuge under cover, usually in small depressions or in a burrow. Its name, racerunner, is derived from its incredible speed during the heat of the day (up to 32.2 km/h [20 mph]).

Population status: Populations of this lizard appear to be stable on Sanibel and Captiva islands in upland areas and the beach.

Threats: With the cessation of most development on Sanibel and Captiva islands, this species does not appear to be threatened by any manmade activity at this time. It is prey for birds, mammals, and other reptiles.

Comments: In 1959 this speedy lizard was locally abundant on the intermittent open sand habitat common along the entire Gulf Beach Ridge Zone and in a few interior sites. The scattered interior sites would have totaled less than 50 hectares in 1960, a time prior to encroachment into this habitat by exotic plants, which resulted in loss of racerunner habitat. This unique habitat of the Mid-Island Ridge Zone and the limited primary habitat that supported *Aspidoscelis* were initially created and maintained over time by the abundant gopher tortoise population and the rather arid environment of the higher ridges. Grazing by the tortoises helped sustain the grassy habitat, which was interspersed with open unvegetated patches of the typical Sanibel Island substrate—a mix of sand and fine seashell particles and coarse grasses on which the tortoises foraged. This provided a habitat niche the racerunner prefers.

Significant changes occurred in the 1970s because of real estate development and the encroachment of exotic and invasive plants. At first, this lizard colonized the new habitat, but its population declined as bare sand was sodded and landscaped or pest plants invaded and eliminated the desirable habitat on vacant subdivision lots.

It was amazing how quickly this lizard located the new habitat and began a pioneer colony in the freshly scraped and disturbed areas. The racerunner would suddenly appear in subdivision construction sites a considerable distance from the nearest established colonies. With the build-out of Sanibel Island, dwelling-unit density changes are unlikely because of a strict land development code that has been in place since 1976. The land use plan has been fostered by the COS governing body, and additional land-clearing for large subdivisions is not likely in the future. The six-lined racerunner may continue to survive on this island in small viable populations in the limited pioneer vegetation community that is part of the Gulf Beach Ridge Zone.

Basiliscus vittatus Wiegmann, 1828

Brown basilisk (Passively introduced)

An adult male brown basilisk. The peculiar dorsal crests, and the ability of this lizard to avoid predators by running across the surface of open water, set it apart from others in the Western Hemisphere. This tropical species is a recent addition to the herpetofauna of Sanibel and Captiva islands. *Daniel Parker*

Other common names: Northern brown basilisk, common basilisk, striped basilisk, "Jesus Christ lizard."

Similar species: Some iguanas and other tropical lizards are somewhat similar in appearance to this species, but none has characteristics that resemble the unique fin-like dorsal crests of this genus.

General range of species: *Basiliscus vittatus* is a native of northern South America and Central America where it ranges into Central Mexico. It was introduced into Southeast Florida in the early 1970s and quickly naturalized. Distributed via the nursery-plant trade, the brown basilisk is now considered common in Dade and southern Broward counties. It is known to be expanding its range in Southwest Florida; there are a few isolated populations in South Fort Myers in Lee County and Naples in Collier County.

Island distribution: Known on Sanibel and Captiva islands from one locality. See range map below.

Preferred habitat type(s): Dense vegetation along streams or canals is prime habitat for this lizard.

Size: The brown basilisk reaches a maximum length of 80 cm (31.5 in).

Color: Large adults are brown. Juveniles and small adults have white or yellow stripes on the face and laterally on the upper body. In the large adult the tail and rear legs may reveal dark crossbands until old age (6 to 8 years of age).

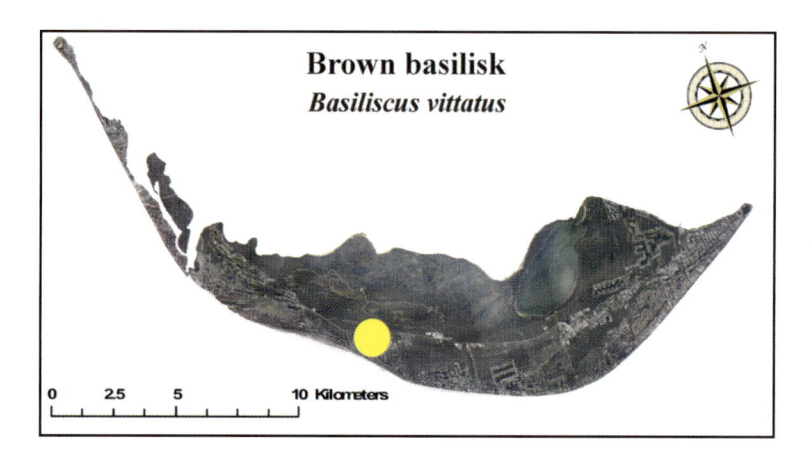

Range Map 12. Collection site of *Basiliscus vittatus* on Sanibel Island in 2013.

Other characteristics: Both sexes have three prominent crests: on the rear of the head, a vertebral crest on the body, and another on the top of the tail. These crests are much larger and more distinctive in the male. Its hind feet have weblike flaps that aid the lizard when it runs across the surface of still water. Young basilisks can usually dash on top of the water up to 18.2 m (60 ft) without sinking.

Diet: Insects are the major part of this lizard's diet but it is known to consume fruit and small vertebrates.

Reproduction: The brown basilisk deposits two to 18 eggs, between five and eight times a year. These average 21.0 mm (.82 in) in length and 13.7 mm (.54 in) in diameter. The eggs hatch after a three-month incubation period. Neonate basilisks are 15.2 cm (6 in) long.

Life history: The brown basilisk is a denizen of densely vegetated habitat associated with water. It basks on trees or brush close to water into which it can leap when threatened. A basilisk will regularly dive into water to escape threats, but it is best known for its unique ability to run across the surface of the water.

Population status: Dynamic populations occur in Southeast Florida, and this lizard is spreading in the ideal habitat of South Florida. A lone individual found DOR in early 2013 is the only representative of this species thus far documented on the islands.

Threats: Since this is an exotic species, we have not considered the threats to its existence in Florida.

Comments: In its Central and South American home range the brown basilisk is commonly called the Jesus Christ lizard. This is because of its ability literally to run rapidly across the surface of still waters on its hind feet without sinking. Younger lizards can speed across the water for longer distances than the adult because the heavier adult sinks earlier.

On 30 April 2013, Lee County Deputy Sheriff Michael Sawicki discovered the remains of an adult brown basilisk on the road in the Gulf Pines subdivision on western Sanibel Island. He documented this with photographs and sent them to Sanibel Island resident and author Charles Sobczak. Sobczak forwarded them to Lechowicz for positive identification of the dead lizard. At this writing it is too early to ascertain if this exotic lizard has established itself on the barrier islands. This incident is likely a result of a stowaway lizard(s) or egg(s) that arrived by way of plants or mulch that were brought to Sanibel Island from either Broward or Dade County, Florida.

Gekko gecko (Linnaeus, 1758)
Tokay gecko (Actively introduced)

When approached, the tokay gecko often assumes an intimidating defensive posture. This lizard is capable of inflicting painful bites, but as in the vast majority of lizards, its bite is harmless. *Bill Love*

Other common names: None.

Similar species: *Hemidactylus sp.* (house gecko).

General range of species: *Gekko gecko* is found throughout Southeast Asia including the Philippines, Indonesia, Bangladesh, India, and New Guinea. It has been introduced into Florida, Texas, Hawaii, Belize, and the Caribbean.

Island distribution: Found sporadically on Sanibel and Captiva islands in disturbed areas, usually near homes or buildings.

Preferred habitat type(s): The tokay gecko prefers forested areas, steep cliffs, and human developments.

Size: A medium-sized lizard, 20 to 30 cm (adults are usually 7 to 11 in long); the largest recorded (a male) measured 38 cm (15 in) in length.

Color: The tokay gecko is blue or gray with red-hued spots on the body.

Other characteristics: *Gekko gecko* is considered the second largest gecko in the world[32]. The tokay gecko has very large eyes with slit-like pupils and large toe pads that enable this agile species to climb up walls and across ceilings.

Diet: The tokay gecko is mostly an insect eater but will also eat amphibians, reptiles, and invertebrates.

Reproduction: The tokay gecko deposits multiple clutches of one or two eggs. It demonstrates parental care over its eggs until they hatch.

Call: The mating and territory defense call of the tokay gecko sounds like a high pitched "*to-kay*" or "*uh-oh.*"

Life history: The tokay gecko is an introduced, nonnative species in Florida. This exotic nocturnal lizard is sometimes released in or near human dwellings because of the large amount of insects they can eat and help control. In this case, its releasers are targeting prey such as cockroaches.

Population status: The abundance of the tokay on Sanibel and Captiva islands does not seem to be increasing. It is common locally as a result of intentional releases, but it does not appear to be aggressively invasive.

Threats: Since the tokay gecko is an exotic species, the threats to its existence in Florida have not been considered. This species is nocturnal, and the islands have no native nocturnal lizards for it to eat. It consumes the exotic, *Hemidactylus* sp. (house gecko), which shares its nocturnal habitat, and it is large enough to eat most native bats, but we have no evidence it does so.

Comments: The tokay gecko was originally released by reptile fanciers at residential locations on the eastern and middle sections of Sanibel Island beginning in the mid-1970s. Weakened captive specimens were purchased at discounted prices from amphibian and reptile dealers in Fort Myers and then liberated on the barrier islands. This Asiatic species seems to be firmly established, and the population is well represented on the island.

Hemidactylus garnotii Duméril and Bibron, 1836
Indo-Pacific house gecko (Passively introduced)

Normally active at night where it is regularly observed on the exterior of buildings, the Indo-Pacific house geckos appears much lighter in the dark than the specimen pictured above in daylight. In fact, it appears to be almost ghost-like, and the white-hued lizard seems to be here one moment and gone the next because of its almost uncanny ability to hide in tiny crevices. This species comes out at twilight and searches for night-active insects on which it preys. *Daniel Parker*

Other common names: House gecko.

Similar species: Wood slave (Tropical house gecko), brown anole.

General range of species: *Hemidactylus garnotii* is from Southeast Asia and India. It is established in Florida, Hawaii, and the Bahamas.

Island distribution: The Indo-Pacific house gecko is found throughout Sanibel and Captiva islands in small areas. Its range on the islands has diminished over the past 20 years.

Preferred habitat type(s): In its native range, this nocturnal lizard is found in trees, under bark, and in crevices. In Florida, it is associated with manmade structures. It hides in the cracks of buildings and in debris piles.

Size: A small lizard, about 10.2 to 13.5 cm in length (adults are usually 4 to 5.5 in long). This gecko's maximum recorded length is 13.3 cm (5.25 in).

Color: *Hemidactylus garnotii* is brown to gray; its body appears translucent at night.

Other characteristics: This gecko has large eyes to see prey at night. It has well-developed toe pads for climbing at all angles. Unlike other Florida *Hemidactylus* species, *Hemidactylus garnotii* has smooth skin.

Diet: The Indo-Pacific house gecko eats insects primarily, but will also eat other small invertebrates.

Reproduction: This species reproduces parthenogenetically, meaning it procreates asexually. This is basically an all-female species in which the unfertilized eggs develop into a clone of the adult. Eggs are deposited in pairs in crevices, under logs, in cracks in buildings, or under various types of debris.

Life history: The Indo-Pacific house gecko has spread all over the world through cargo shipments. It is almost always associated with human habitation, living on the inside and outside of people's homes or businesses. It often finds a crevice near a light that is illuminated at night so it can easily catch insects that are attracted to the light.

Population status: The Indo-Pacific house gecko appears to be much less common on Sanibel and Captiva islands than 20 years ago, but it still occurs in small pockets on Sanibel Island.

Range Map 13. Initial collection site of *Hemidactylus garnotii* on Sanibel Island in 1971.

Threats: Since this is an exotic species, the threats to its existence in Florida have not been considered. It is not known to cause any harm to the natural environment or native species. It is simply occupying an unoccupied niche (a native, nocturnal insect-eating lizard is absent). Therefore, *Hemidactylus garnotii* does not compete with any other lizards except for other exotic *Hemidactylus* species.

Comments: *Hemidactylus garnotii* was first observed by George Weymouth on Sanibel Island at a public shower facility at the Periwinkle Park and Campground. Weymouth brought this previously unknown population to LeBuff's attention in 1971 (G. Weymouth, pers. comm.). Not long after this, herpetologist Jack McCoy coincidentally visited the island, where he collected specimens and later published his collection data (McCoy 1972). This gecko was destined for early success, and within 10 years the species was distributed throughout Sanibel and Captiva islands.

The more recent introduction of another gecko, the wood slave (formerly called the tropical house gecko, *Hemidactylus mabouia*), has adversely affected *Hemidactylus garnotii*. The wood slave has replaced the Indo-Pacific gecko in many areas in South Florida, including Sanibel and Captiva islands. Most of the geckos now seen at night on the islands are wood slaves.

Hemidactylus mabouia (Moreau de Jonnès, 1818)
Wood slave (Passively introduced)

The wood slave gecko is a relatively recent arrival that has successfully colonized Sanibel and Captiva islands. There is evidence that this lizard is outcompeting the well-entrenched, 40-year-old population of the Indo-Pacific house gecko on the barrier islands. *Daniel Parker*

Other common names: House gecko.

Similar species: Indo-Pacific house gecko (*Hemidactylus garnotii*), juvenile tokay gecko (*Gekko gecko*).

General range of species: *Hemidactylus mabouia* is found naturally in sub-Saharan Africa. It has spread to North, South, and Central America due to its habit of stowing away on exported freight.

Island distribution: Found throughout Sanibel and Captiva islands in natural and manmade environments.

Preferred habitat type(s): The wood slave (formerly known by the common name, tropical gecko) is found naturally in tropical forested areas but has thrived in and around human dwellings. This exotic, nocturnal lizard is often found under bark, behind fastened objects on walls, and near outside lights.

Size: A small lizard, 11 to 12.7 cm (adults are usually 4.3 to 5 in long), reaching a maximum recorded length of 14 cm (5.5 in).

Color: The wood slave appears cream colored or white at night, but during the day it is gray to brown.

Other characteristics: The wood slave gecko has large eyes with slanted pupils and large toe pads that enable it to climb walls and cross ceilings.

Diet: It eats insects, scorpions, spiders, and sometimes other lizards.

Reproduction: The female *Hemidactylus mabouia* usually deposits one to three eggs in tight crevices.

Call: This gecko is vocal and can make a sound resembling the word "gecko," which is often repeated three times.

Life history: The wood slave gecko is an exotic species in Florida and has been very successful at colonizing new areas. It may gather communally under cover by day.

Population status: Populations of *Hemidactylus mabouia* on the islands and the mainland are stable and more than likely spreading. This gecko was first discovered in the U.S. in the Florida Keys in 1990 and has quickly spread to most of South Florida, displacing other exotic *Hemidactylus*.

Threats: Since the wood slave is an exotic species, the threats to its existence in Florida have not been considered. This gecko poses little threat to any native species. Since there are no native geckos in Southwest Florida, this lizard does not compete with native species. However, there are other nonnative species on Sanibel and Captiva islands, such as the Indo-Pacific house gecko, with which it competes at night.

Comments: *Hemidactylus mabouia* has become established on the South Florida mainland and the Florida Keys. Lechowicz (2009) included this species as a recent arrival that has gained a foothold on Sanibel Island. Undoubtedly this lizard was transported to the barrier islands through the nursery plant trade. Soon *Hemidactylus mabouia* will become well established on the island, just as it has succeeded in other parts of Florida.

Iguana iguana (Linnaeus, 1758)
Green iguana (Actively introduced)

Adult green iguanas develop complex morphological characters, including a combination of bizarre scalation and coloration. These features make their identification easy. *Daniel Parker*

Other common names: None.

Similar species: None.

General range of species: *Iguana iguana* is found naturally in parts of Mexico, Central America, South America, and the Caribbean. It has been introduced into Florida, Texas, Hawaii, the U.S. Virgin Islands, and several of the Bahama Islands.

Island distribution: Found in residential areas and mangrove habitats on Sanibel and Captiva islands.

Preferred habitat type(s): The green iguana is arboreal. It spends most of its time in the tree canopy, but will bask and forage on the ground. It prefers tree canopies near water so it can jump into the water when a predator is near.

Size: A large lizard, 1.3 to 1.5 m (4.3 to 4.9 ft), with a record length of 2.0 m (6.6 ft).

Color: *Iguana iguana* can be various colors depending on their age, sex, and the weather.

Juveniles are a bright lime green. The male tends to show much red or orange coloring when it is hot from basking. This lizard can be almost completely black or gray when the air temperature is cold but the sun is shining; the dark color helps it absorb heat from the sun.

Range Map 14. Location of the first documented *Iguana iguana* on Sanibel.

Other characteristics: The green iguana has one row of spines down its back and tail. A large dewlap helps regulate temperature and is displayed for territorial and courtship purposes.

Diet: *Iguana iguana* is primarily herbivorous (sometimes referred to as a folivore or leaf eater). It eats fruits and flowers, as well.

Reproduction: The female green iguana lays 20 to 71 eggs that require 10 to 15 weeks to incubate. On Sanibel Island, it usually lays the eggs in a burrow.

Life history: Most of this iguana's time from juvenile to adulthood is spent in trees. Young green iguanas tend to stay together in small groups. The green iguana is one of the most popular pet lizards in the world, and for that reason it is hunted for export and raised on breeding farms. It is also a delicacy in many of its home countries and therefore is under much hunting pressure. It is an invasive, exotic species in Florida.

Population status: The population has decreased substantially since the freeze of 2010 and because of the COS Removal Project, but the species is still present on the island and will regain numbers with time. Cooler than average winter temperatures causes mortality in this species and reduces its numbers.

Threats: Since the green iguana is nonnative to the U.S., threats to its existence are not considered. The most significant predators of adult iguanas on Sanibel Island are probably American alligators, and dogs and cats (including bobcats). Juveniles are eaten by birds of prey (e.g., red-shouldered hawk).

Comments: *Iguana iguana* was first reported on Sanibel Island in 1970. The first specimen reported to JNDDNWR was an adult using an abandoned gopher tortoise burrow in a vacant lot at the southeast corner of Periwinkle Way and Beach Road on the island's east end. The lizard was presumed to be an escapee or a specimen that had been intentionally

A juvenile green iguana. Small, immature specimens were very popular in the pet trade during the 1980s and 1990s, but they soon became hard-to-handle adults, and their owners grew tired of them. Many former pet iguanas were released into the wild in Southwest Florida, including Sanibel Island—long a repository for unwanted wildlife pets. Daniel Parker

released by a reptile fancier. Owners who make such releases may have the well-being of the animals at heart, but releasing them on an island known for its renowned sanctuary atmosphere is still not warranted.

In 2007 the COS implemented a Nile Monitor and Green Iguana Management Plan with an initial budget of $55,000 ($20,000 came from a USFWS grant). By the third quarter of 2009, a contractor had removed 1,607 iguanas from Sanibel Island at a cost of $42,905 (J. Isom, pers. comm.).

Most of the eradication efforts were conducted in residential neighborhoods where iguana populations were the densest. It appears to prefer back yards where people plant colorful exotic flowers and plants and have gardens with vegetables and fruit trees, which the lizard eats. It is especially fond of *Hibiscus* sp. The green iguana is seldom found in natural habitats on the island, with the exception of the Mangrove Zone. It is occasionally seen running through the interior conservation lands along the Sanibel River corridor but rarely stays there very long. An exception is along the periphery of protected land adjacent to human development. The restored natural lands owned by JNDDNWR, SCCF, and COS on the island probably could support colonies of the green iguana, especially near the Sanibel River, but the food source is not as appealing as it is in well-landscaped residential yards with ample colorful food items.

Leiocephalus carinatus armouri (Linnaeus, 1758)
Northern curly-tailed lizard (Passively introduced)

Well-named for its peculiar ability to curl its non-prehensile tail vertically, this curly-tailed lizard was photographed with its appendage in that unique position. It is usually curled when the lizard is excited or when males are trying to attract females during courtship. When it is at rest, the tail is usually straight as is typical for most lizards. *Bill Love*

Other common names: Curlytail lizard.

Similar species: None on Sanibel or Captiva Island.

General range of species: *Leiocephalus carinatus armouri* is found naturally on a few of the Bahama Islands (Grand Bahama and the Abaco Islands). It was released in Palm Beach in the 1940s and has expanded its range since. In Florida, it is found on the southeast coast and at a few localities along the west coast. This species is spreading quickly.

Island distribution: The northern curly-tailed lizard was first found on Captiva Island in 2008 and at one locality on Sanibel Island in 2011. The Captiva Island colony is expanding. As of 2013, only two specimens have been found on Sanibel Island.

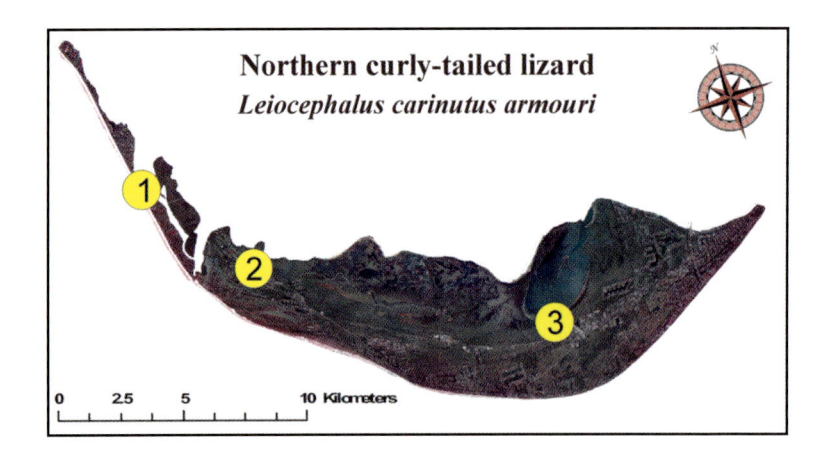

Range Map 15. Localities of collection points of the northern curly-tailed lizard. (**1**) Captiva (2008), (**2**) Sanibel (2011), (**3**) Sanibel (April 2012).

Preferred habitat type (s): The curly-tailed lizard occurs in upland areas such as beaches or dunes. It prefers rocky or sandy areas.

Size: The adult ranges from 18 to 33 cm (7 to 13 in).

Color: The curly-tailed lizard is brown or gray on the dorsal side and yellow or white on the ventral side.

Other characteristics: *Leiocephalus carinatus armouri* has pointed, keeled scales on the dorsal side. Its body has either spots or stripes, and it has a speckled throat.

Diet: Omnivorous, the curly-tailed lizard has been observed taking discarded fruit, but it is mostly an insect eater. This lizard also eats other invertebrates and small reptiles.

Reproduction: The curly-tailed lizard breeds in the spring and deposits two to six eggs in early to mid-summer.

Life history: *Leiocephalus carinatus armouri* is an invasive, exotic species in Florida. It is very successful at colonizing new areas, and thus is referred to as invasive. This diurnal species spends its day hunting for insects and lizards on the ground, but will climb into trees when fleeing a predator or chasing prey. It curls and/or twitches its tail as a defensive mechanism to distract a predator.

Population status: The population of *Leiocephalus carinatus armouri* is growing rapidly on Captiva Island. As expected, it has worked its way south, and the two records on Sanibel Island probably are only the first of many more to come.

Threats: Since the northern curly-tailed lizard is an exotic species, the threats to its existence in Florida have not been considered. The threat it poses to native fauna is unknown. It may eat the native green anole (*Anolis carolinensis*) from time to time. It does eat the exotic brown anole (*Anolis sagrei*), which prefers habitats near the ground, thus placing it in visual range of the curly-tailed lizard.

Comments: The northern curly-tailed lizard was first documented on Captiva Island at 'Tween Waters Inn on 11 July 2008. Sam Landry, the SCCF Captiva Sea Turtle Coordinator, made the initial observation, and a specimen was captured by Lechowicz. The first record of this lizard on Sanibel Island was on 22 August 2011 when Lechowicz verified a photograph taken by Sanctuary Golf Course superintendent Kyle Sweet.

The northern curly-tailed lizard likely was introduced by the nursery trade to Captiva Island immediately after Hurricane Charley in 2004.

Ophisaurus ventralis (Linnaeus, 1766)
Eastern glass lizard

The eastern glass lizard had not been collected or observed on Sanibel Island in more than 50 years until 2012. This completely limbless lizard differs from snakes in several ways. It has eyelids and external ear openings, but snakes have neither. About two-thirds of this lizard's total length is tail, and the tail is easily broken. The tails of most lizards regenerate, but those of snakes do not. *Daniel Parker*

Other common names: Glass lizard, glass snake.

Similar species: Two additional species of *Ophisaurus* occur on the nearby mainland, but neither is known to occur on the barrier islands.

General range of species: *Ophisaurus ventralis* is a southeastern species that ranges from southeast Louisiana throughout all of Florida, minus some barrier islands, and northward up the coastal plain to southeast Virginia.

Island distribution: *Ophisaurus ventralis* was known from only one individual found on eastern Sanibel Island in 1959. It was considered extirpated until an injured glass lizard was discovered on Island Inn Road in 2012.

Range Map 16. Sites of where the two individuals of *Ophisaurus ventralis* were collected on Sanibel Island in 1959 (**1**) and 2012 (**2**).

Preferred habitat type(s): The eastern glass lizard is typically an upland species but is also found around moist habitats. This lizard inhabits sandy, hardwood hammocks in South Florida but is found in pine flatwoods and grasslands elsewhere.

Size: A large lizard, averaging 56 to 108 cm (adults are usually 22 to 42 in long). The maximum SVL recorded for this lizard is 30.6 cm (12 in). The tail often accounts for more than two-thirds of this lizard's maximum total length of 108 cm (42.5 in).

Color: The eastern glass lizard is primarily tan to brown with some longitudinal stripes. Some older individuals may appear green. It has white markings on the posterior sections of the scutes, as well as on the neck.

Other characteristics: This legless lizard superficially resembles a snake. Unlike snakes, it has both ear openings and eyelids that enable it to close its eyes. It can lose its tail and regenerate a new one like most other lizards, but unlike snakes.

Diet: The glass lizard eats mostly insects and other arthropods, although it has been documented to eat the eggs of reptiles and ground-nesting birds.

Reproduction: *Ophisaurus ventralis* is a summer breeder. The female regularly deposits about a half-dozen eggs under cover such as logs, boards, and trash. She watches over the nest until the eggs hatch.

Life history: The eastern glass lizard is a diurnal predator. It seeks refuge under the sand surface and ambushes prey that crosses its path. It is called a glass lizard because its tail tends to break rather easily, often at the first sign of trouble. It is not uncommon to see Florida specimens with dark regenerated tails.

Population status: Glass lizard specimens have been seen twice on Sanibel Island, although there were 53 years between observations. During those years, the glass lizard was assumed to be extirpated on the barrier islands.

Threats: The eastern glass lizard population is threatened by road traffic and land development.

Comments: The dried remains of an eastern glass lizard were found on the pavement of Periwinkle Way on the island's east end by LeBuff in 1959. The habitat where this glass lizard was found has since been completely modified through construction of the dredge-and-fill canal-front home sites of Shell Harbor Subdivision. LeBuff always assumed that if a population of glass lizards was established 50 years ago, then the secretive lizard could still exist on Sanibel Island but had simply gone undetected. This assumption was verified on 30 July 2012, when an eastern glass lizard was struck by an automobile on Island Inn Road near the center of Sanibel Island. The individual was taken to CROW, but did not survive its injuries.

Plestiodon inexpectatus (Taylor, 1932)
Southeastern five-lined skink

By the time it has attained full adult characteristics, the southeastern five-lined skink's stripes become obscure and the head becomes red-hued. *Daniel Parker*

Other common names: Five-lined skink, scorpion, blue-tailed skink (juveniles and subadults).

Similar species: Little brown skink.

General range of species: *Plestiodon inexpectatus* is found in the southeastern U.S. from the Mississippi River in Mississippi and Louisiana, north to Maryland and including the entire state of Florida.

Island distribution: *Plestiodon inexpectatus* occurs sporadically on Sanibel and Captiva islands in natural and manmade habitats.

Preferred habitat type(s): The southeastern five-lined skink is found in dry woodland and dune habitats. This lizard prefers places with many layers of habitat, such as woodpiles or even garbage piles. It can be found in terrestrial and arboreal habitats.

Size: A moderately sized lizard, adults are usually 12.5 to 21.5 cm (4.9 to 8 in) long. The record length for this species is 21.6 cm (8.5 in).

Color: The southeastern five-lined skink is black, gray, or brown with five dull white lines running lengthwise along the dorsal side of the body. The stripes become more orange as they approach the head. For very old males the stripes are ill-defined on the dark background, and parts of the head may become a rusty-red color.

Juvenile southeastern five-lined skinks have a bright blue tail. This coloration changes over time, and the blue-pigmented tail color is usually lost by the time an individual skink reaches adulthood.
Daniel Parker

Other characteristics: *Plestiodon inexpectatus* has a bright blue or purple tail as a juvenile. This lizard's body is smooth, unlike most lizards in the region. The male head tends to have more orange on it than the female's head. The body of the adult female tends to be browner than the male.

Diet: This lizard eats insects, various invertebrates, amphibians, and reptiles.

Reproduction: *Plestiodon inexpectatus* deposits four to 14 eggs in early summer. The female skink stays with the eggs during incubation.

Life history: The southeastern five-lined skink is a diurnal species that spends its day scurrying among the leaf litter and low brush to bask and search for insects. It is able to climb up the sides of walls. The tail of this animal breaks off rather quickly if handled or harassed.

Population status: This skink's status on Sanibel Island and the mainland is stable.

Threats: *Plestiodon inexpectatus* does not have any major threats on Sanibel or Captiva islands. It thrives in island neighborhoods, as well as in various natural habitats. It is eaten by snakes, birds, and occasionally mammals.

Comments: *Plestiodon inexpectatus* has been a common member of the herpetofauna of Sanibel Island through the last half-century, and this relatively secretive species continues to occur throughout the island. On some of the mid-island ridge subsets, its abundance may have been reduced in recent years because of the controlled burn programs.

Scincella lateralis (Say *in* James, 1823)
Little brown skink

The little brown skink is a secretive species that spends most of its life concealed under leaf litter. Once very common on the barrier islands, it is now seldom encountered during herpetological surveys. Elimination of undesirable Australian pines because of positive habitat restoration work has directly contributed to the decline in this species' population on the barrier islands. During controlled burns, the incineration of the dense leaf litter that had accumulated beneath the standing but dead pines drastically decreased herpetofaunal habitat. *Daniel Parker*

Other common names: Ground skink.

Similar species: Juvenile southeastern five-lined skinks, although the coloration of the two species is entirely different.

General range of species: *Scincella lateralis* is found from New Jersey to Kansas and throughout the southeastern U.S.

Island distribution: Currently found primarily in the interior open woodlands of Sanibel and possibly Captiva islands.

Preferred habitat type(s): Leaf litter, debris piles, and rotting logs in woodlands, usually near water. This lizard has a preference for loose soils in which it can burrow at the first sign of distress.

Size: Small, 7.4 to 14.5 cm (3 to 3.5 in), with a maximum recorded length of 13 cm (5 in).

Color: The little brown skink is brown to almost black on the dorsal side with a longitudinal black stripe on each side of the upper body. The ventral side is usually creamy white or yellow.

Other characteristics: *Scincella lateralis* is a very small skink with a proportionately long tail and short limbs.

Diet: This lizard eats small insects and invertebrates.

Reproduction: The female deposits eggs (1 to 6) in moist soils or crevices and can deposit two to three clutches per year. This lizard does not guard its nest, unlike the southeastern five-lined skink.

Life history: *Scincella lateralis* is a terrestrial species and rarely climbs trees. It mostly lives in leaf litter or debris piles in damp vegetated areas. It loses its tail quite easily when harassed by predators or when caught and held.

Population status: *Scincella lateralis* was quite common on Sanibel Island in 1959 but is rarely encountered today, although populations may be stable. In recent years, this skink has been observed only on conservation lands near the Sanibel River. The presence and status of *Scincella* on Captiva Island are unknown.

Threats: The little brown skink is eaten by many vertebrates such as housecats and several invertebrates. Other than predation by wildlife and mortality caused by infrequent fires, it has no other known threats at this time. On Sanibel Island, this skink is a favored food of coralsnakes (*Micrurus fulvius*) and is also eaten by several other snake species.

Comments: In 1959 this small secretive skink was regularly found in the leaf litter of stands of exotic Australian pines, *Casuarina equisetifolia*. This was the only habitat in which it could be collected on a regular basis, and it was common beneath Australian pines at the SILS and similar habitats throughout the island. With the distribution of *Casuarina* now considerably more restricted because of control efforts since 1980, coupled with the tree's widespread removal following Hurricane Charley in 2004, the islandwide population of this diminutive lizard has apparently declined.

Varanus niloticus (Linnaeus *in* Hasselquist, 1758)
Nile monitor (Assumed self-introduced)

A juvenile Nile monitor. Great effort is being made by wildlife and land managers to keep this voracious predator off Sanibel and Captiva islands. This invasive species must not be allowed to colonize the barrier islands. *Bill Love*

Other common names: Goanna.

Similar species: Juvenile alligators.

General range of species: *Varanus niloticus* is an exotic species found naturally in central and southern Africa.

Island distribution: *Varanus niloticus* is known from three documented sightings on Sanibel Island. One sighting was on the far west end of the island, one was centrally located, and the other was near the east end.

Preferred habitat type(s): The Nile monitor inhabits riparian areas. In Florida, it prefers the banks of ditches and canals.

Size: A very large lizard reaching up to 270 cm (9 ft), but generally about 183 cm (72 in) in length, including the tail.

Color: The adult Nile monitor is primarily gray-brown with green or yellow spots arranged in bands. The juvenile is more vividly colored, being mostly black with dark yellow or green-hued blotches.

Other characteristics: The Nile monitor has a keen sense of smell. It has an enlarged forked tongue that helps it detect the scent of prey. Large, strong claws help it dig and climb and allows it to rip into the meat it is eating. Its tail is very powerful and can be whipped at predators, including humans, with incredible strength.

Diet: The Nile monitor eats fish, amphibians, reptiles, small mammals, and invertebrates (that is, almost anything that moves), as well as the eggs of any of the oviparous species. Across Pine Island Sound, in Cape Coral, Todd Campbell from the University of Tampa conducted a study in 2005 to find out what monitors were eating there. He found that cockroaches made up the largest percentage of the diet, although burrowing owl feathers were found in one animal during necropsy.

Reproduction: The female *Varanus niloticus* digs an underground nest and deposits 10 to 60 eggs, which typically take up to nine months to hatch. At hatching, a Nile monitor is already 15.2 to 20.3 cm (6 to 8 in) long. The female has been known to return to the nest to help her young reach the surface at hatching.

Life history: The Nile monitor is a diurnal species. It is strictly carnivorous and very good at catching prey. It spends its life foraging near water, and it will jump into the water at the slightest disturbance. The Nile monitor can hold its breath for up to an hour depending on the water temperature. It can climb trees and quickly disappear into active or abandoned animal burrows.

Population status: The Nile monitor has not been documented on Sanibel or Captiva islands since 2008. It is known only from three documented records from 2006-2008 (a photograph, a video, and a dead animal). We currently have no evidence of a breeding population on Sanibel or Captiva islands.

Range Map 17. The location of the first documented *Varanus niloticus* on Sanibel Island, 12 August 2005.

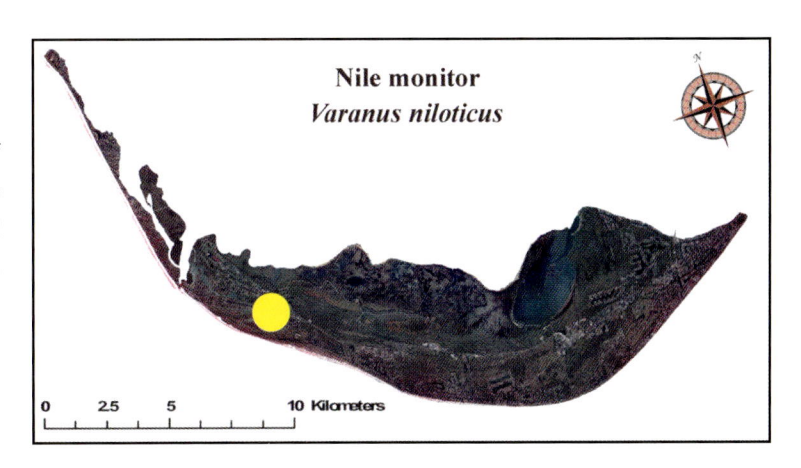

Threats: There are no threats to the Nile monitor in Florida, except humans. Because of the monitor's proclivity for water, alligators can attack them. The Nile monitor, however, is a threat to many animals, and would be especially so if it were ever to become established on Sanibel or Captiva islands. The threat to the eggs of sea and freshwater turtles, as well as ground-nesting birds and other egg-laying reptiles, could be catastrophic. The monitor is a threat to ornamental fish such as goldfish and koi that are kept in outdoor ponds. It is doubtful that a Nile monitor would attack a small dog or adult cat, and it is unreasonable to think it would attack a person.

Comments: *Varanus niloticus* was documented on Sanibel Island in 2005. Its presence was first substantiated by photographs and video documentation, and later by finding a dead specimen. This species breeds on the mainland in Cape Coral, and recently has been found on Pine Island, across Pine Island Sound from Sanibel Island. Its introduction to Sanibel Island may have been facilitated by its propensity to swim across water bodies, but it could also be that someone made a poor decision and actively introduced this lizard to Sanibel Island. Perhaps all three documented observations on Sanibel Island were of the same animal that was moving around; no Nile monitors have been observed on either Sanibel or Captiva islands since the dead lizard was found. A necropsy of the dead Sanibel Island Nile monitor was performed by Lechowicz in 2008. All that was found in the lizard's gut were a few small egg fragments. This animal had obviously not eaten in a while, and possibly suffered from an illness prior to its demise. It had no external wounds or scarring.

Despite policies to curtail its establishment, *Varanus niloticus* will likely become successfully established on our barrier islands, although human pressure to eradicate may ensure that populations here will never become as large as those of *Iguana iguana*. We likely will experience the effects of their presence as they negatively impact native amphibians, turtles, alligators, their eggs and hatchlings, and reptiles, birds, and small mammals.

~ORDER SQUAMATA—THE SNAKES~

Coluber constrictor priapus Dunn and Wood, 1939

Southern black racer

An adult southern black racer. This snake is sometimes confused with the eastern indigo snake, but it usually lacks the iridescence of the indigo, and the racer has a bright white chin, is smaller, and more slender. *Daniel Parker*

Other common names: Black snake, racer, black rat snake.

Similar species: Eastern indigo snake.

General range of species: The southern black racer is found throughout the southeastern U.S. There are isolated populations in west Texas and a subspecies that ranges into Mexico from South Texas.

Island distribution: *Coluber constrictor priapus* is found throughout Sanibel and Captiva islands in all habitats.

Preferred habitat type(s): The southern black racer is a habitat generalist. It occurs in most Florida ecosystems and is common in both natural and disturbed habitats. It is not secretive and is often seen in residential neighborhoods.

Size: A relatively large species of snake, the southern black racer can attain a length of 50 to 142 cm (adults are usually 20 to 56 in) in length. The record length is 182 cm (72 in).

Color: The adult southern black racer is primarily black, bluish, or gray dorsally. The venter is white, and this snake has a red-hued iris.

Other characteristics: *Coluber constrictor priapus* as a neonate appears completely different from an adult. Hatchlings are white or gray ventrally with red-toned blotches or bands. The venter has pink spots on a white background. As they age, the black racer becomes more and more gray and is completely black at about two years of age.

Diet: The racer eats lizards, snakes, rodents, amphibians, and insects.

Reproduction: The southern black racer is oviparous, meaning its eggs develop and hatch outside the mother's body. It deposits from six to 20 eggs in the late spring or summer. The eggs are elongated and have a granular surface, with an average size of 17.9 mm by 44.7 mm (0.7 by 1.75 in).

Hatchling southern black racer. Well patterned at hatching, this species darkens with age. *Chris Lechowicz*

Life history: The southern black racer is diurnal. It is primarily terrestrial, but will quickly climb trees or shrubs to escape predators or catch prey. The species is seen frequently and will enter open garages during cool weather. Its name derives from its nervousness and speed. The species name *constrictor* is a misnomer. It overpowers and devours its prey without using constriction.

Population status: *Coluber constrictor priapus* is common throughout Sanibel and Captiva islands.

Threats: There are no serious threats to southern black racers on the islands. On Sanibel and Captiva islands this snake is eaten by birds and mammals. It was once prey of the now extirpated indigo snake.

Comments: *Coluber constrictor priapus* remains one of the most abundant snakes on Sanibel and Captiva islands. This species occurs in upland habitats and frequents subdivisions where it is often an active and visible predator of the brown anole (*Anolis sagrei*) and the Cuban treefrog (*Osteopilus septentrionalis*).

In 2002 and 2003, Lechowicz interviewed many long-time residents on Sanibel Island about the abundance of the southern black racer. All respondents thought that the racer was more common in the 1990s and 2000s than in the 1960s through 1980s. They suggested that the presence of the eastern indigo snake, a well-known snake predator, kept the racer population low.

Coluber flagellum flagellum (Shaw, 1802)
Eastern coachwhip

An adult eastern coachwhip photographed at Fannie's Preserve, a Sanibel Island tract of land owned and managed by SCCF. The snake's common name was derived from the appearance of the species' posterior, which resembles a braided whip. *Chris Lechowicz*

Other common names: Coachwhip snake, coachwhip, whip snake.

Similar species: Coachwhip juveniles can be easily confused with those of the black racer and the ratsnake.

General range of species: This is a snake of the southeastern states. The coachwhip occurs from North Carolina to Florida, and west to eastern Texas. Its range extends to parts of Oklahoma and the southeastern corner of Kansas.

Island distribution: In 1959, the coachwhip was distributed throughout the major ridges on Sanibel Island, but populations are now confined to the Gulf Beach Ridge Zone ecosystem and a few upland habitats that have been restored.

Preferred habitat type(s): The coachwhip is usually associated with open scrub, grassy areas, and pine flatwoods. On Sanibel and Captiva islands, remnant populations of this snake usually occupy habitats that are part of dry upland systems (e.g., the Gulf Beach Ridge Zone and any of the few remaining open, thinly vegetated localities).

Size: *Coluber flagellum* is one of North America's longest snakes. At hatching, the coachwhip averages about 355 mm (14 in), and adults average 107 to 152 cm (42 to 60 in) in length. The record length for this species is 259 cm (102 in).

Color: The coloration of the eastern coachwhip is variable across the breadth of its geographical range. In Southwest Florida, including Sanibel and Captiva, the adult *Coluber flagellum* is basically two-toned. The majority of its body color is a light tan, appearing almost white in older adults, while the head and neck region is generally a dark brown. This contrasting color is responsible for its common name; the head and neck resemble the handle, and the body looks like a braided whip; hence, coachwhip.

Other characteristics: *Coluber flagellum* is North America's fastest snake. Published accounts report that the coachwhip is able to attain speeds of 13 km/h (10 mph), although this is far slower than the erroneous tales embedded in American folklore claiming this species can outrun a horse.

Diet: The coachwhip preys on small mammals, ground-nesting birds and their eggs, lizards, and other snakes.

Reproduction: The coachwhip is an egg-laying species. The female deposits 12 to 16 elliptical, granular-surfaced eggs (Ortenburger 1928), usually under vegetative debris or rotting logs where the level of moisture is conducive to their development. On average, the eggs measure 17 by 63 mm (0.67 by 2.48 in) and hatch following a 70-day incubation period.

Life history: Well-patterned at hatching with crossbands, the young coachwhip grows quickly, and the distinct juvenile body markings are lost over a two-year period. With further growth, the ground color of this species' body transitions to an overall tan color in Southwest Florida. Following the subadult period, the neck and head darken as maturity is attained. The coachwhip is considered a terrestrial species, but individuals occasionally move into low trees or bushes to pursue prey, to bask, or to escape danger.

A juvenile, post-hatchling, eastern coachwhip. The blotches of pigment on the neck and thin crossbands on the body soon disappear as the snake grows. Daniel Parker

Population status: Coachwhip populations in the region have been reduced over time because of land development and fragmentation of habitat. Outside of large parks, reserves, and refuges, this species will likely continue to decline.

Threats: The coachwhip, because it is large and obvious, is often killed by hunters and construction workers. In some parts of Florida, many die while crossing roads. Visibility increases the coachwhip's vulnerability, and it becomes an easy target for uncaring snake-killing motorists. Fragmentation of habitat and land development also continues to adversely impact the survival of this species everywhere it occurs.

Comments: Once common in the Gulf Beach Ridge Zone and on some of the open mid-island ridge of Sanibel Island, the speedy coachwhip still occupies the remnants of its former habitat in this ecosystem.

One of the most fascinating observations the senior author ever witnessed was a life-and-death struggle between a very large adult coachwhip and an indigo snake of equal length. The battle unfolded on the entrance road leading to the then refuge-managed SILS on Point Ybel, Sanibel Island. The two snakes were situated in the center of the shell-surfaced road, and the indigo snake had the coachwhip firmly in its jaws. It was attempting to hold its prey to the ground with its heavier body as it energetically worked the coachwhip's head ever closer to its mouth. The coachwhip flailed about, but the indigo persisted and soon gained the head. It began to consume its prey until the tail of the coachwhip disappeared. Afterwards, the indigo snake made some stretching bodily adjustments to compensate for the size of its meal, and it slowly disappeared into the dense West Indian vegetation.

A mark-recapture study performed by Lechowicz in 2006-2008 in the Gulf Beach Ridge Zone on Sanibel Island showed that *Coluber flagellum* was the second-most abundant snake in that zone. The southern black racer was the most abundant. One male coachwhip ranged more than 9.6 km (6 mi) in the narrow dune vegetation along the beach over a period of three months. During that time, Lechowicz observed a coachwhip consuming a marsh rabbit at a west-end study site.

Crotalus adamanteus Palisot de Beauvois, 1799
Eastern diamond-backed rattlesnake

This defensively coiled, medium-size *Crotalus adamanteus* was photographed in 1988 on Sanibel Island's JNDDNWR on a remote shell ridge where it was partly concealed in the shadow of a seagrape tree. *Charles LeBuff*

Other common names: Diamondback rattler, rattler, diamondback.

Similar species: No other southeastern rattlesnakes are similar.

General range of species: *Crotalus adamanteus* inhabits the southeastern U.S., primarily coastal plain between southeastern North Carolina and Florida and the parishes of eastern Louisiana. This species is found in all Florida counties, including the Florida Keys.

Island distribution: The eastern diamond-backed rattlesnake encountered on Sanibel Island has usually been associated with the Gulf Beach Ridge ecosystem. The majority of historical observations were restricted to the eastern end of Sanibel Island. The eastern diamond-backed rattlesnake occurred on Captiva Island until the late 20th century. This snake was never recorded on Upper Captiva, although the species currently exists on Cayo Costa to the north. There have been no recent observations of this snake, and it is considered by Lechowicz to be extirpated on Sanibel and Captiva islands.

Preferred habitat type(s): On the adjacent mainland, *Crotalus adamanteus* occupies pine flatwoods/palmetto communities or elevated ridge systems known as "pine islands" in the Big Cypress Basin. Unfortunately, its preferred habitat is the first to go when major land development occurs. On the islands, the ecological vitality of the Gulf Beach Ridge Zone has been drastically altered over the past 50 years and is approaching housing-unit build-out. Should colonizers of this species reach the islands, there is little suitable habitat remaining to support a viable population of this nationally jeopardized snake.

Size: The eastern diamondback is the largest of the rattlesnakes and the largest venomous snake in North America, with an average length of 167.6 cm (65 in). The record total length is 243.8 cm (96 in), but today wild eastern diamond-backed rattlesnakes of such proportions are probably few in number.

Color: The variable basic color of this snake is usually some shade of olive brown, which in Southwest Florida can be correlated with a specimen's habitat preferences. A snake from open palmetto prairie and sand pine scrub areas appears lighter hued than the much darker specimens that inhabit denser, shadier oak scrub and flatwoods forests. This handsome snake's famous diamond pattern consists of dark brown to black rhombus-shaped blotches that are well defined because of the double row of cream- to yellow-colored scales that contact and surround them. The eastern diamondback's head is uniquely marked with lines on its face. Other eastern species of rattlesnakes do not have this facial coloring, although several species in the western U.S. do.

Other characteristics: The telltale rattle at the end of the tail is a characteristic that immediately identifies members of the genus *Crotalus*. The eastern diamond-backed rattlesnake is an extremely venomous snake. This species is responsible for many of the snakebites in Florida. Some of these are serious enough to result in the amputation of fingers, toes, and even limbs; in the most severe cases, death can result. There are no records or memories of envenomation by a venomous snake on the islands.

Diet: Small mammals are the prey of choice for this large snake, and subadult *Crotalus* are fond of the hispid cotton rat (*Sigmodon hispidus*) and other rodents. The adult diamondback on the mainland regularly consumes eastern cottontail rabbits (*Sylvilagus floridanus*) as its main prey, as well as eastern gray squirrels (*Sciurus carolinensis*). The cottontail rabbit is not known to occur on Sanibel and Captiva islands, and it has always been assumed this snake consumes cotton rats and Florida marsh rabbits (*Sylvilagus palustris paludicola*) on the islands.

Reproduction: Fertilization in snakes is internal for all but a limited number of asexual species. Many crotaline species, including eastern diamondbacks, go through a courtship ritual between the sexes. The eastern diamond-backed rattlesnake is ovoviviparous, meaning it gives birth to living young. Large females in some populations will produce more than a dozen young that at birth average 370 mm (15 in) in length. This snake is potently venomous even at birth. The neonate has a pre-rattle terminal base called a "button" at the end of its tail, from which the famous rattle grows.

Life history: Each time a rattlesnake molts, or sheds its skin, as part of its normal growth it adds another segment to its rattle. In good habitats with sufficient prey, this snake grows fast and may shed up to 12 times a year. So, contrary to the old wives' tale, you cannot determine the age of a rattlesnake by counting the segments of its rattle.

In Florida, the eastern diamond-back rattlesnake prefers habitats close to escape refuges such as gopher tortoise burrows, crevices or holes in cap rock, or dense sheltering clumps of saw palmetto. This rattlesnake is somewhat solitary and secretive. When cornered or shut off from escape, the snake will defensively coil while simultaneously vibrating its tail. The noisy rattle announces the snake's nervous presence, in effect warning the intruder to stay clear. When given an opportunity, this snake will attempt to avoid a confrontation and retreat.

Population status: Many herpetologists share concern for the viability of the eastern diamond-backed rattlesnake's survival. It has disappeared from most suburban habitats where it once flourished prior to the huge real estate development and agricultural lands clearing that destroyed much of wild Florida in the latter part of the last century. In rural areas, it is impulsively killed when it comes in contact with humans because of the species' maligned reputation and the public's misconception of the snake's value to ecological communities. Outside of the larger and more remote parks and forests, this snake is declining in Florida. The diamondback continues to be killed by employees of lease-holders and concessionaires on public lands.

Threats: High-density residential and commercial development, the destruction of vast habitats for farming, ranching, and citrus groves, and an ever-increasing grid of highways and roads threaten the survival of the eastern diamond-backed rattlesnake. Automobile traffic alone has decimated the population of this snake in Southwest Florida. Roadways and subdivisions have seriously fragmented the habitat necessary for the survival of *Crotalus adamanteus* in the state.

Another major issue has been the indiscriminate gassing of gopher tortoise burrows by "professional" snake hunters and contestants in "rattlesnake roundups," such as those in southern Georgia and Alabama. Collectors use gas to drive out and collect diamondbacks seeking refuge in tortoise burrows. Use of gaseous chemicals is now prohibited as a collection method throughout Florida.

Insertion of raw gasoline into gopher tortoise burrows is an example of man's cruelty toward nature and was directly responsible for the deaths of countless thousands of gopher tortoises, gopher frogs, eastern indigo snakes, eastern diamond-backed rattlesnakes, and other endangered species that sought refuge in the tortoise burrows.

After a long absence, the Florida bobcat (*Lynx rufus floridanus*) returned to the Sanibel Island ecosystems in the 1970s. This nonindigenous feline quickly competed with this rattlesnake and the eastern indigo snake for marsh rabbits and small rodents. Since its reintroduction, the bobcat's growing numbers may have adversely impacted the populations of both snakes by reducing their prey base. Bobcats may have simply become another of the many causative factors that have contributed to the extirpation of both snakes on our barrier islands.

Comments: In 1959, the eastern diamond-backed rattlesnake was well represented on parts of Sanibel Island. The individuals usually encountered in the field, those found and reported by refuge visitors at headquarters, or those reported on rare occasions by telephone (after private telephones became available) were all large adults. The individuals collected were usually measured and then released at a selected relocation site, and averaged 1.5 m (48 in) in total length. Neonate or even subadult specimens were never reported or encountered on Sanibel Island, although a large rattlesnake killed on Donax Street circa 1956 (D. Hiers, pers. comm.) by a homeowner was found to be a female containing 13 near full-term embryos.

Live *Crotalus* were usually discovered crossing roads that transected the wetland uplands, but some were found DOR on the Mid-Island Ridge Zone corridor. Specimens were occasionally discovered on the gulf and bay beaches. In 1964, one large individual was found dead on C span of the Sanibel Causeway.

A review of LeBuff's personal copies of his official USFWS Weekly Narrative records disclose instances of encounters between people and several adult *Crotalus adamanteus* on eastern Sanibel Island. These include a specimen killed at the fishing pier at Point Ybel in September 1961 (LeBuff 1998); another adult killed by Sanibel icon Clarence Rutland in his Periwinkle Way yard on 22 October 1964; and an adult killed on the gulf beach ridge immediately west of the SILS by James Williams at Jim's Shell Shop on the Shanahan property on 8 August 1965. In July 1974 the late Allen Nave, founder of Nave Plumbing, live-captured a large adult rattlesnake on Main Street, Sanibel Island. He brought it to the refuge office, and LeBuff later released the snake on an isolated upland ridge in the BT a distance away from public-use areas.

The largest *Crotalus adamanteus* LeBuff ever personally measured (rostral to base of rattle) on Sanibel Island was 221 cm (87 in) in total length. LeBuff was present when a Lee County Deputy Sheriff shot and killed this specimen at the intersection of West Gulf Drive and Rabbit Road on Sanibel Island in 1960 (LeBuff 1998), and he collected the carcass and measured it.

The eastern diamond-backed rattlesnake population on Sanibel Island has decreased alarmingly and today is virtually nonexistent on the islands. Concurrent decreased abundance of this species on the adjacent mainland has reduced the opportunities for new recruits to make the swim to Sanibel Island.

With seasonal vehicle traffic so heavy today, few living creatures make it safely across any of the roads on Sanibel or Captiva islands. LeBuff last observed and photographed a midsized eastern diamond-backed rattlesnake on JNDDNWR in 1988 (above photograph). The specimen was situated on an inland ridge, part of the northern primary subset of the Mid-Island Ridge Zone that is dominated by typical West Indian hardwoods. A major Lee County Electric Cooperative elevated power line crosses the western part of the refuge over a wide easement from Pine Island Sound, crossing this ridge and providing easy human ingress into the habitat.

The last known eastern diamond-backed rattlesnake on Sanibel Island was collected on West Gulf Drive by Rob Loflin, COS, and relocated to the JNDDNWR in 1996. Numerous unconfirmed reports have been received from residents and tourists, but no island biologist, ecologist, or herpetologist has been able to document the snake's presence on Sanibel Island since 1996.

Diadophis punctatus punctatus (Linnaeus, 1766)
Southern ring-necked snake

The southern ring-necked snake is positioned to reveal both its dorsal and ventral coloration. The ventral surface of this snake's tail is generally redder than that of the body's venter. When defending itself, this small harmless snake exposes the underside of its tail to reveal the red color in hopes a predator will be rebuffed by the bright warning color. *Daniel Parker*

Other common names: Ringneck snake, ringneck, worm snake.

Similar species: Peninsula brown snake.

General range of species: The southern ring-necked snake is found throughout the eastern U.S. and southeastern Canada from the Florida Keys to Nova Scotia, and as far west as eastern Minnesota in the north and Arizona in the south. Separated subspecies are found from California to Washington and into Idaho, Wyoming, and Nevada.

Island distribution: The ring-necked snake is found throughout the islands in woodlands and around residences.

Preferred habitat type(s): The preferred habitat of *Diadophis p. punctatus* is woodlands and hammocks. It is often found in backyards and debris piles near human habitation.

Size: This is a small snake, 25 to 38 cm (10 to 15 in) long, that has attained a record total length of 48.2 cm (18.87 in).

Color: The southern ring-necked snake is gray, black, or blue-toned with an incomplete orange or yellow neck band. The venter can be a variety of colors from yellow, orange, or red to a combination of those.

Other characteristics: *Diadophis p. punctatus* has a row of black spots down the center of the venter. The neck ring usually is incomplete in this subspecies; it neither meets nor connects dorsally and is separated by a few dark scales. The female is larger than the male.

Diet: The ring-necked snake eats a variety of vertebrates and invertebrates. Prey consists mostly of worms, slugs, insects, and some amphibians and reptiles.

Reproduction: *Diadophis p. punctatus* is oviparous, and a clutch will vary from one to 10 elliptical eggs. On average, these measure 9 by 20 mm (0.35 by 0.78 in). A female ring-necked snake can produce more than one clutch of eggs per year.

Life history: The southern ring-necked snake is a mildly venomous snake. The venom does not affect humans but is quite useful in immobilizing small prey. The snake also uses constriction to subdue its prey. Considered social by some herpetologists, the ring-necked snake is often found with others of its species.

Population status: The southern ring-necked snake is common on the islands in most wooded habitats, even in disturbed areas.

Threats: This species is eaten by mammals, birds, and other snakes. There is considerable habitat for it on Sanibel and Captiva islands, so it is in no immediate danger.

Comments: *Diadophis p. punctatus* is abundant on ridges and their margins throughout Sanibel Island. The snake is regularly observed around residential landscaping near dusk, and often comes out of concealment during summer torrential rain events. In September 1959, a specimen collected on Sanibel Island was submitted to the OSU collection, and catalogued as R-1992.

Drymarchon couperi (Holbrook, 1842)
Eastern indigo snake

In sunlight, the eastern indigo snake is highly iridescent, reflecting a rainbow-like blue-black coloration. The color of their chin and throat is very variable. The head of the specimen pictured above is pigmented similarly to individuals of this species that formerly occurred on Sanibel and Captiva islands. Because of its large size and usually gentle demeanor, the indigo snake was once prized in the pet trade. *Daniel Parker*

Other common names: Indigo snake, gopher snake, blue gopher snake.

Similar species: Southern black racer, eastern coachwhip snake.

General range of species: *Drymarchon couperi* was historically found in southeastern Mississippi, southern Alabama, the coastal plain of extreme southern South Carolina[33], eastern Georgia, and throughout the state of Florida.

Island distribution: The indigo snake is close to extirpation from Sanibel and Captiva islands, and was considered so until an adult was captured on Captiva Island on 20 September 2012. This was the first individual to be documented on that island since 1988.

Preferred habitat type(s): The preferred habitat of the indigo snake is open grassy uplands near gopher tortoise burrows, which it uses as refuges in the winter and during fires. The indigo snake prefers places with nearby water sources. It tends to frequent more mesic (moist) areas in the summer.

Size: *Drymarchon couperi* is North America's largest native nonvenomous snake, and adults average 100 to 180 cm (39 to 70.9 in) in total length. The maximum recorded total length for this species is 262.9 cm (103.5 in).

Color: The eastern indigo snake is iridescent blue or black, and its ventral side can be as dark or even grayish. Many individuals have red chins and throats, although those on Sanibel and Captiva islands did not have much red compared with other populations.

Other characteristics: The indigo snake is characterized by its large size and shiny blue-black coloration. When in the sunlight the large, mostly smooth scales of this snake produce a unique iridescence.

Diet: The eastern indigo snake will eat almost any live animal it can get in its mouth, with the exception of many invertebrates.

Reproduction: *Drymarchon couperi* is usually a winter breeder. In Southwest Florida, this species conducts a complex courtship ritual, and the male ranges widely through his individual territory when sexually active. Unfortunately, this exposes it to road traffic. Copulation occurs in the winter months, and females produce a clutch of four to 14 large, elongated eggs that measure 31.75 by 50.8 mm (1.25 by 2 in) on average. Eggs are deposited in moist vegetative debris (e.g., beneath rotting logs or in animal burrows that suit the environmental needs of incubation) in April or early May. The 45.7-cm (18-in) young emerge after a 90- to 120-day developmental period. In some populations, hatchling indigo snakes have a defined light and dark pattern. All the neonate *Drymarchon couperi* the senior author has observed on Sanibel Island, however, were similar to the adult coloration—that is, solid blue-black, without any discernible pattern.

Life history: *Drymarchon couperi* is a diurnal "racer-like" snake that has a very large home range and requires expansive tracts of unbisected land (no roads) in order to be successful. This snake has a reputation of being very docile, and most specimens will rarely bite or resist when picked up. This is why the species was once so popular in the pet trade. LeBuff, however, has memories of a few painful bites inflicted on his forearms, hands, and fingers by some defensive indigo snakes he casually picked up when he found them on JNDDNWR or elsewhere on Sanibel Island. The indigo has been listed as a Threatened Species by USFWS since 1978 and was protected by FFWCC prior to that.

The indigo snake is not a constrictor; it simply suppresses and holds its prey in place with its mouth, in conjunction with its body weight and its muscular pressure, as it ingests the prey.

Population status: The eastern indigo snake was assumed recently extirpated on Sanibel and Captiva islands until 2012.

Threats: The indigo snake is imperiled by highway mortality, incidental collecting, and malicious killing. On the islands juveniles were typically eaten by birds, mammals, and southern black racers.

Comments: The eastern indigo, *Drymarchon couperi* (formerly *D. corais couperi*), was once a common species on Sanibel and Captiva islands—that is, until a few years after the completion of the Sanibel Causeway. This roadway brought staggering numbers of vehicles to the islands, and by 1972 traffic had increased to gridlock proportions during ensuing winter tourist seasons. The 1972 Annual Narrative Report for JNDDNWR stated that *Drymarchon* *couperi* began its decline on Sanibel Island that year. An indigo snake, a specimen considered to be the last known on Sanibel Island, was killed when a bicyclist ran over it in February 1999 on the refuge's popular Indigo Trail[34]. It was taken to CROW for veterinary treatment but had been mortally injured and died a day later. SCCF and JNDDNWR have conducted many surveys and projects since then to locate this snake on Sanibel Island, but to no avail. Residents and tourists occasionally report them, but these sightings are undoubtedly misidentified southern black racers.

In 1959, J. N. "Ding" Darling (pers. comm., LeBuff 1998) recognized the serious threat that the increase in vehicular traffic posed for the indigo snake and other terrestrial forms of island wildlife, even before the ferry service ceased operation.

An unfortunate event occurred on Captiva Island during Thanksgiving week, 2012. An eastern indigo snake was intentionally killed on the northern end of the island. The person who killed the snake claimed it was harassing some visitors and he took it upon himself to dispatch the snake. Over a week later, the remains of this snake were found by USFWS and brought to Lechowicz for positive identification. The individual contained the internal PIT-tag installed by Lechowicz as part of the SCCF Pine Island Sound Eastern Indigo Snake Project. This microchip proved that this individual was indeed the only known indigo snake from Captiva since 1988 (Its collection is referenced above). The rediscovery of an indigo snake on Captiva was a popular story on the TV news, island newspapers, and the internet. Hopefully, the killer of this snake will be prosecuted to the full extent of the law and the penalty for his act will be severe. The outcome may serve as a lesson to others to just leave these rare, harmless animals alone. (As of October 2013 this case is in the hands of the FFWCC and the outcome is pending.)

Lampropeltis elapsoides (Holbrook, 1838)
Scarlet kingsnake (Passively introduced)

A beautiful specimen of the scarlet kingsnake. Its red-tipped nose and the fact that the yellow crossbands are contained within black crossbands, thus separating the red from the yellow, are keys to the proper identification of this coral snake "mimic." *Daniel Parker*

Other common names: Scarlet king, kingsnake.

Similar species: Eastern coralsnake, Florida scarletsnake.

General range of species: The scarlet kingsnake is a southeastern snake found from eastern Louisiana, throughout Florida, and as far north as Virginia. Isolated populations occur in western Kentucky.

Island distribution: The occurrence of *Lampropeltis elapsoides* on Sanibel Island is known from only one specimen found in an island residence.

Preferred habitat type(s): The scarlet kingsnake is found in wet to dry prairie, pinelands, or forests. It hides in small, tight crevices in logs, under bark, and in rock piles.

Size: This is a small snake ranging from 35 to 50 cm (adults are usually 14 to 20 in) in length. The maximum recorded length for a scarlet kingsnake is 68.6 cm (27 in).

Color: This tricolored snake has alternating bands of red, black, and yellow. Red always borders black and not yellow. The bands are continuous around the snakes' body. The tip of the head is red. Some specimens have white bands instead of yellow bands.

Other characteristics: *Lampropeltis elapsoides* is a small kingsnake compared with the other U.S. *Lampropeltis* species. Until recently this snake was considered a subspecies of the milksnake but has now been recognized as a full species.

Diet: Its diet consists of lizards, snakes, and rodents.

Reproduction: The scarlet kingsnake is oviparous. It deposits between two and nine eggs in the late spring or summer. Eggs are elliptical and average 11 by 25 mm (0.43 by 0.98 in).

Life history: The scarlet kingsnake is often mistaken for the coralsnake because of its superficially similar coloration. As a result, many are killed when encountered. This secretive snake is rarely seen during the day, but it may be observed crossing the road on warm rainy nights.

Population status: The scarlet kingsnake is not currently known to occur on Sanibel or Captiva islands.

Threats: If the scarlet kingsnake were to be found on the islands, it would not be a serious threat to any native animal. It is eaten by birds, mammals, and snakes.

Range Map 18. Collection site of *Lampropeltis elapsoides* on Sanibel Island in 1977.

Comments: On Sanibel Island, *Lampropeltis elapsoides* is known from only one specimen that was collected inside a residence on West Gulf Drive in 1977 by George Weymouth (Campbell 1978; G. Weymouth, pers. comm.). This individual later escaped from captivity on the eastern end of the island.

Other individual scarlet kingsnakes have undoubtedly reached Sanibel Island by crossing the causeway along with fill dirt, potting soil, or as stowaways in landscape plant containers. Individuals are likely to be observed in the future.

Micrurus fulvius (Linnaeus, 1766)
Harlequin coralsnake

An adult harlequin coralsnake. This snake's yellow and red crossbands are adjacent and the tip of the nose is black, behind which is a broad yellow band. This venomous snake was once among the most common species on Sanibel Island. *Bill Love*

Other common names: Eastern coral snake, coral snake, harlequin snake.

Similar species: The scarlet kingsnake (*Lampropeltis elapsoides*) and the Florida scarletsnake (*Cemophora coccinea*[35]) are similar in color, although their band pattern and color sequences are different. In the coralsnake the red and yellow bands touch, but in the other two they do not.

General range of species: The harlequin coralsnake occurs throughout the lowlands of the southeastern U.S., basically following the coastal plain from southern North Carolina to north central Louisiana, excluding the Mississippi Delta parishes. There is a localized population in central Alabama. This coralsnake is found in all Florida counties, including the upper Keys. Another species, the Texas coralsnake (*Micrurus tener*), occurs in southern Arkansas, western Louisiana, eastern and central Texas, and northern Mexico.

Island distribution: This coralsnake was abundant historically in all the ridge systems on Sanibel Island and was observed with regularity. Specimens were frequently observed in the margins of the more upland swales—that is, those not regularly tidally flooded, including narrow sections of the red mangrove forest that connect to and parallel the mid-island ridge.

Preferred habitat type(s): On Sanibel Island, the coralsnake has always preferred the dry habitats available on the higher ridges, especially those dominated by cabbage palms and West Indian hardwoods. The surface of those elevated ridges is typically blanketed with a thick substrate of fallen cabbage palm fronds. In such areas, the coralsnake lives in microhabitats provided by the often moist, humus-filled pockets at the basal connection of old frond boots that remain attached to the trees. The Mid-Island Ridge Zone probably supported the greatest numbers of this snake. In the early 1960s, specimens were commonly found DOR on Periwinkle Way and Sanibel-Captiva Road. Specimens were occasionally observed on the surface in part of the dry, old mangrove system north of the mid-island ridge on JNDDNWR and on the COS bicycle path that follows Sanibel-Captiva Road along the national wildlife refuge boundary. Adult coralsnakes were regularly seen above ground at other various favorable habitats on the refuge.

Size: A common misconception among laymen is that the coralsnake is small, but this is not the case when its maximum size is considered. At hatching, this snake averages 17.8 cm (7 in) in total length. The adult snake averages 60 cm (23.6 in) in length, and males grow longer than females. The published record length for *Micrurus fulvius* is 120.7 cm (47.5 in), from a specimen found in North-Central Florida. In 1956, LeBuff collected a live coralsnake in a residential yard on Coconut Drive, in Fort Myers, Lee County, Florida, that measured 111.1 cm (43.75 in), which may be the record length for this species in Southwest Florida. For many years, after preservation, this impressive specimen was housed in the privately owned museum collection at Everglades Wonder Gardens, Bonita Springs, Florida.

Color: The bright color pattern of the coralsnake is set in rhyme: "Red touch yellow, kill a fellow; red touch black, friend of Jack." This is correct; the wide red and narrow yellow rings are separated from contacting each other by those that are black and also wide. Typically in this region of Florida, the snake's body is completely ringed with bright bands of red, yellow, and black except the snake's tail, which is ringed in black and yellow only. In some specimens (such as the individual in the photograph), there are often a large number of black scales within the color field of the red rings. One of this species' key distinguishing color characteristics is the color of its snout; from its tip to the side of the head at a point behind the eyes, the snout is black. This black nose and the adjacent, contacting, wide yellow band on the posterior section of the snake's head can be relied on to identify normal-colored coralsnakes.

Other characteristics: *Micrurus fulvius* is extremely venomous, but bites to humans are rare. Unlike the pit vipers (snakes of the genera *Agkistrodon*, *Crotalus*, and *Sistrurus*), which are outfitted with retractable fangs that fold up toward the roof of the snake's mouth when relaxed, the coralsnake's fangs are fixed in place and permanently erect. Rather than a rapid strike that injects a large venom load deep into the fang punctures, as done by members of the three genera above, the coralsnake sometimes chews at a bite site to better inject its venom. Most severe bites to humans have been inflicted on fingers and toes or the soft tissue between them, but this snake's small mouth can also get a hold on larger body parts. Conversely, the rattlesnake can inject venom by a simple low-energy bite, although the conduit of major envenomation accompanies a forceful strike. The venom of the coralsnake is a neurotoxin and one of the most dangerous snake venoms, but this snake's bites are rarely fatal to humans because of the poor delivery system.

Diet: The coralsnake preys on small snakes and ground-dwelling lizards; it is also known to be cannibalistic, often eating its own kind in captivity. Young rodents, nestling ground-nesting birds, terrestrial and fossorial (burrowing) lizards, and frogs are also part of this snake's diet.

Reproduction: On our barrier islands, the coralsnake is sexually active in the spring, typically in March and April. This snake is oviparous, meaning it lays eggs. Females produce from two to a dozen eggs from May through July. These extremely elongated eggs average 12.7 by 38.1 mm (0.5 by 1.5 in) and hatch in approximately 80 days. Just as the pit vipers are venomous at birth, so is the coralsnake at hatching.

Life history: With a sufficient food supply, the coralsnake grows rapidly. It is semi-fossorial and considered to be a secretive snake. On our islands, the coralsnake prefers habitats with either loose soils or those that are blanketed with dense leaf litter to a substrate of deep soft humus. Except for contact with conspecifics during mating, the coralsnake is usually a territorial loner. There are accounts in the literature that describe the antagonistic relationships individual coralsnakes often have with one another when they meet.

Population status: The population of coralsnakes on Sanibel and Captiva islands is now stressed and apparently declining due to a variety of factors. Development, road traffic, ecosystem failure, and adverse land management have contributed to the decline. Elsewhere, the species is safe within large parks, forests, and wildlife refuges. The last documented coralsnake on Sanibel Island was photographed in 2002 on the BT. It is likely that a few *Micrurus* remain on the island, but no DOR specimens have been collected, and no reliable undocumented sightings have been reported to Lechowicz.

Threats: Land development, road traffic, and controlled burns may take a toll on the few coralsnakes remaining on the islands.

Comments: This highly venomous elapid was once a common species on Sanibel Island. At night during the warmer months, and especially during rainstorms, individual coralsnakes were frequently observed crossing island roads and bike paths; however, it is currently considered rare because of a variety of factors (Lechowicz 2009). Coralsnakes were regularly seen moving on the surface of the thin layer of leaf litter in the West Indian hardwood hammocks of the Mid-Island Ridge Zone, and even in the thick leaf litter of old landlocked strands of red mangrove that have not been flooded by saltwater for at least 87 years. Drift fences installed by Lechowicz and JNDDNWR staff have failed to trap the species on the refuge or on lands managed by SCCF. A combined search by the authors in former coralsnake habitats in 2011 failed to find this species.

As a management strategy, JNDDNWR and SCCF have regularly conducted controlled burns in recent decades. Some of these fires included combustion of leaf litter on those ridges dominated by the palm *Sabal palmetto* or West Indian hardwoods, or on the strands of landlocked, nontidal red mangroves. Although we have no evidence other than a reduced population of coralsnakes, the application of fire to these habitats may have inadvertently caused coralsnake mortality, and these fires may have temporarily but effectively reduced both this snake's population and its food supply, which have been slow to recover.

There are other factors to be considered, and we do not suggest that fire alone is responsible for decline of the herpetofauna. Fire ants (*Solenopsis* sp.) are known to prey on ground and semi-fossorial lizards and snakes at an increasing level throughout Florida, including our barrier islands.

Another factor leading to the reduction of the coralsnake population and possibly that of other leaf litter-dependent species (*Plestiodon*, *Scincella*, *Diadophis*, and *Storeria*) on which coralsnakes prey is the complete removal of stands of the introduced Australian pine, *Casuarina equisetifolia*, that had become established on sections of "conservation lands."

During LeBuff's tenure on the refuge, coralsnakes were regularly observed on one of the westernmost tracts, a popular public-use area known as Shell Mound Trail. At this site, the abundance of coralsnakes was considered to be a safety issue for members of the Youth Conservation Corps when they were assigned trail/boardwalk maintenance duties. Each summer from 1976-1986, the teenagers and their supervisor regularly encountered adult coralsnakes. At that time, the cleared trail of this elevated boardwalk meandered over the ground surface and was not completely raised as it encircles an extensive prehistoric Calusa Indian site. This midden mound connects to the major ridge that extends 3.5 km (2.17 mi) to the east where it connects to the Mid-Island Ridge Zone, which Sanibel-Captiva Road follows. At the ridge's western end, typical West Indian hardwoods dominate; near its eastern terminus, this ridge supports Sanibel Island's only natural live oak (*Quercus virginiana*) community. East of the Shell Mound Trail, this ridge is closed to public use, and only a 30.5-m (100-ft)-wide power-line right-of-way and a series of maintained fire lines break the vegetative continuity of the system.

Farther east, close to the refuge's easternmost boundary, is a tract of land that has been designated as part of the National Wilderness System, as has most of the mangrove forest of the refuge that is situated north of Wildlife Drive. The tract is former farmland, and 1920s-era furrows were still visible on the land's surface in 1990. After much of Sanibel Island was inundated by a major hurricane's storm surge in 1926, a factor which contributed to a collapse of the vigorous farming industry on the island, this land slowly made a vegetative transition to a community dominated by West Indian hardwoods.

Then the nonindigenous *Casuarina equisetifolia* gained access after seeds were transported to the site by the tidal surge of Hurricane Donna in 1960. This exotic tree, which by 1960 was growing prolifically along the beachfront toe of the Gulf Beach Ridge Zone, had become established on the outer beach by the hurricanes of the 1940s and 1950s. The nearest source of seeds in those decades that began this exotic's beach invasion probably was a double row of cultivated *Casuarina* growing along a gulf-front road between Casa

Ybel and Island Inn. The Australian pine is a fast grower in Florida. This tree is known to grow over 3.05 m (10 ft) a year under cultivation (Morton 1980). The habitat beneath the mature trees was transformed over time. The first consequence of this invasion was the gross loss of desirable native vegetation because of shading, followed by the rapid build-up of *Casuarina* leaf litter, which provided habitat for small vertebrates. Consecutive fire applications in such habitats elsewhere on the refuge eliminated the leaf litter, and the dependent herpetofaunal species decreased in abundance, possibly to an unknown pre-*Casuarina* population because of removal of the non-indigenous leaf litter.

By 1980, these trees dominated the canopy and another exotic, the Brazilian pepper (*Schinus terebinthifolius*), crowded out most of the understory. Beginning in 1981, both noxious trees were systematically killed as part of a management strategy on this tract by basal application of the herbicides 2, 4-D, or Garlon 4. Over a period of a few weeks, the former herbicide was injected into the *Casuarina* and in a few months all of these trees were dead. During routine pest plant-killing incursions into this tract by LeBuff, adult coralsnakes were regularly observed as they actively moved about on the surface of the leaf litter. Based on those observations, this particular system supported a robust population of *Micrurus fulvius*. This parcel of land and a 0.4-km (0.25-mi)-long manmade spoil ridge at its western boundary were not included in early drift fence trapping on JNDDNWR. A small section of this tract was inspected by the authors in March 2011. Its character had changed since the removal of the *Casuarina*, and no coralsnakes were observed; the area with old farm furrows was not searched.

Nerodia clarkii compressicauda Kennicott, 1860
Mangrove saltmarsh watersnake

A group of mangrove saltmarsh watersnakes. Each represents one of the myriad color variations this watersnake is known for. These are all *Nerodia clarkii compressicauda*. *Daniel Parker*

Other common names: Mangrove watersnake, flat-tailed watersnake.

Similar species: The Florida watersnake (*Nerodia fasciata pictiventris*) has some structural and occasional pattern characteristics that are superficially similar to those of the mangrove saltmarsh watersnake. The two are known to hybridize where their habitats overlap on the margins of brackish water. Two different races of saltmarsh snakes occur north of the range of *N. c. compressicauda* on each side of the Florida peninsula.

General range of species: *Nerodia clarkii compressicauda* is confined to the red mangrove-dominated coastal zones of the peninsula of South Florida, the Florida Keys, and northern Cuba. Generally, this snake's restricted range includes Florida's gulf coast from Citrus County south through the Keys and around to the Atlantic coast extending as far north as Brevard County.

Island distribution: This highly aquatic snake occurs throughout the mangrove habitats of the island.

Preferred habitat type(s): This saltmarsh snake is found almost exclusively in the red mangroves and immediately adjacent tidal areas.

Size: A relatively small watersnake, this species averages 76 cm (30 in) in total length. The maximum recorded size for the mangrove saltmarsh watersnake is 93.3 cm (36.75 in).

Color: The color of *Nerodia clarkii compressicauda* is highly variable, and there are two major morphs or color variations with intermediate phases between the two. If any color is normal for this subspecies, it is a green-gray base color with darker body spots and sometimes broken stripes on the neck. Saltmarsh watersnakes observed around the barrier islands are quite variable, ranging from gray to red-orange. Some individuals may be a carrot-colored red that lack spots or striping. The snake's coloration serves as camouflage, allowing it to be well concealed while basking or foraging among the prop roots of the red mangrove.

Other characteristics: The mangrove saltmarsh watersnake was originally known as the flat-tailed watersnake, hence its subspecific name, *compressicauda* (meaning compressed tail). There are other interpretations of the root of this name's origin. Some herpetologists have argued that this subspecies' tail is flattened to facilitate swimming in its aquatic habitat; others were of the opinion that the tail of the type specimen[36] had been crushed and somewhat flattened, so much so it was interpreted as being "oar-" or "paddle-like." Contemporary herpetologists do not consider this snake's tail to be flattened.

Diet: The diet of this watersnake is restricted to small fish.

Reproduction: The mangrove saltmarsh watersnake is ovoviviparous; the female produces litters of living young, and bears up to 25 hatchlings after a 90-day gestation period. It produces far smaller broods than most other closely related species. At birth, neonates average 160 mm (6.3 in) in length, and siblings within a brood may vary widely in coloration and pattern.

Life history: Predation by crabs, birds, and fish take a toll, and the percentage of a brood that survives to maturity is very low. This snake spends its life in the brackish water of the Mangrove Zone where there is little competition from other snake species. An unusual behavior exhibited by *Nerodia clarkii compressicauda* has recently been described: it practices lingual luring—that is, it uses its tongue as an attractant to lure fish (Hansknecht 2008).

Population status: The mangrove saltmarsh watersnake has declined on Sanibel Island in the eastern section of JNDDNWR because of the fragmentation of its habitat along the refuge's eastern boundary. Overall, the population in the central part of the refuge, outside of the refuge impoundments, has remained unchanged since the construction of Wildlife Drive. With state-mandated mangrove protection, this snake's population in the CHES should remain essentially stable unless other environmental factors, such as chemical pollution from riverine discharges, impact the overall ecosystem.

Threats: Mangrove saltmarsh watersnakes on JNDDNWR are considered safe from most human-related perils usually faced by snakes (e.g., road traffic and habitual snake-killing people). The populations in the Clam Bayou and Blind Pass habitats have been negatively impacted by habitat degradation in recent years. Effects of the apparent increase in nutrient pollution in the lower CHES on this snake are presently unknown.

Comments: In the mid-20th century, *Nerodia clarkii compressicauda* was abundant in the red mangrove forest of Sanibel Island. When Dixie Beach Boulevard was built its alignment passed through part of Ladyfinger Lakes, and for a few years this snake often crossed the road by the dozens nightly. Watersnakes on the road were noticeably more abundant during the regular periodic spring tides. Both the red and gray color variants were collected, but the latter predominated. After Sanibel Isles Subdivision was nearly completed and more housing units were built on Woodring Point, the population of watersnakes plummeted because of increased vehicular traffic. Lechowicz still finds *Nerodia clarkii compressicauda* on Dixie Beach Boulevard during and after rainstorms, but in very small numbers. This snake is regularly observed in other areas of the refuge dominated by tidal red mangroves.

Nerodia fasciata pictiventris (Cope, 1895)
Florida watersnake

The adult Florida watersnake is also variable in coloration. This snake's body color may become almost uniformly black with age. *Bill Love*

Other common names: Florida banded watersnake, banded watersnake, water moccasin, moccasin.

Similar species: The venomous Florida cottonmouth is similar in general appearance to this harmless species, yet a number of definitive clues set the two apart. The similarity is best considered when comparing adults of each species. An irritated adult Florida watersnake is exceptionally adept at changing the appearance of its head and body to become more threatening and to appear cottonmouth-like. On these islands, larger specimens of the mangrove saltmarsh watersnake may be confused with this freshwater species.

General range of species: This *Nerodia fasciata* subspecies ranges from extreme southern Georgia into Florida where it is distributed throughout the freshwater habitats of the peninsula.

Island distribution: The Florida watersnake occurs throughout the interior wetlands of Sanibel Island. Occasionally, individuals may even range into brackish water habitats near the mangroves.

Preferred habitat type(s): All permanent freshwater bodies in Florida have populations of this watersnake, which frequents the margins of lakes, ponds, and canals.

Size: This is a large watersnake. Average-sized adults reach 106.7 cm (42 in), and the record size for this snake is 158.8 cm (62.5 in) in total length.

Color: Neonates are brightly marked and are highly variable in coloration, even within the same brood. They normally have red or black crossbands on a ground color of complex grays or tans. As the Florida watersnake ages, its coloring darkens and its blotches tend to disappear as dark hues begin to dominate and obscure the crossband pattern. Some enormous adults are very dark and are best described as black.

Other characteristics: The Florida watersnake is pugnacious, although nonvenomous, and can be difficult to handle. It will bite savagely when handled carelessly. Release of foul-smelling liquid stored in this snake's anal musk glands are another defensive behavior, quickly repelling any animal or human that grabs it. The Florida watersnake is somewhat more tolerant of being handled than other members of the genus, but will readily strike if harassed. Interestingly, *Nerodia* has an anticoagulant in its saliva that causes a somewhat bloody wound due to the reduction of clotting activity from the punctures on its many backward-directed teeth.

Diet: Fish and amphibians constitute the main prey of this species.

Reproduction: Ovoviviparous and prolific best describe the reproductive strategy of this species of *Nerodia*. Brood size ranges from 20 to 30 young, born during the warm months between late spring and the end of summer.

Life history: Within days of birth, the young of the Florida watersnake consume small minnows and grow rapidly. By the time this watersnake reaches maturity, prey preferences have increased to include larger fish and amphibians. At all stages of its life cycle, this snake is preyed upon by a variety of wildlife species. Pig frogs, kingsnakes (where they occur), heron and egrets, white ibis and wood storks, turtles, alligators, and otters continually put pressure on this species' survival.

Population status: *Nerodia fasciata pictiventris* is declining on Sanibel Island because of habitat fragmentation and associated road kills. It has also declined in other regions throughout the Florida peninsula. This snake was so abundant 50 years ago that the population seemed literally inexhaustible.

Threats: Drainage and alteration of freshwater habitat through the years have had adverse effects on some populations of this snake. Fragmentation of habitat because of large subdivisions and their close street grids has resulted in a serious increase in vehicle-caused mortality on heavily traveled roadways. The destruction of this species and other watersnakes in the Everglades region began after completion of the Tamiami Trail (U.S. 41) in 1928. This corridor of wildlife destruction violated the Everglades ecosystem, but it alone wasn't enough for "develop-at-all-cost" politicians who were cozy with real estate speculators. U.S. 41 was followed by Alligator Alley (now Interstate 75) in 1969, which multiplied the environmental impacts to this unique world-class ecosystem. The carnage to snake populations and other wildlife we have witnessed on these highways since their construction is impossible to convey. The sheer density of the number of snakes that died while attempting to cross these two highways for the past 85 years is unfathomable for the herpetologists or laymen of today to comprehend. You had to see it to believe it.

Comments: *Nerodia fasciata pictiventris* is regularly observed in the interior wetlands of Sanibel Island. Prior to 1979, this species ranged into the red mangrove forest to reach freshwater held in the West Impoundment of JNDDNWR (see map, page 64). The frequency of sightings of both live and DOR specimens noticeably declined during the years following the island's permanent connection with the mainland in 1963 because of road traffic that seriously fragmented their habitat. Mortality among large adults is correlated with their frequent attempts to cross roads and the increase in nighttime motor vehicle traffic on those arterial roads spanning the freshwater wetlands.

The western end of Sanibel Island along Sanibel-Captiva Road and between Tarpon Bay Road and the Gulf Ridge Subdivision (a distance of 7.25 km [4.5 mi]) seems to be a barrier between the Florida watersnake and the mangrove saltmarsh watersnake. The wetlands on the north side of Sanibel-Captiva Road are primarily brackish as opposed to the south side of the road, which is part of the freshwater Sanibel Slough wetlands. Lechowicz has documented countless roadkills of both species on Sanibel-Captiva Road but has never documented a live mangrove saltmarsh watersnake on the south side of the road and has only once documented a live Florida watersnake on the north side of the road.

Nerodia floridana (Goff, 1936)
Florida green watersnake

An adult Florida green watersnake. The green hue is retained as the snake ages. This species is locally common in some habitats on the mainland, but only one specimen is known from Sanibel Island. *Bill Love*

Other common names: Green watersnake.

Similar species: Other watersnakes of the genus *Nerodia* are similar to one another, but in appearance this snake is quite different because of its unique coloration.

General range of species: The green watersnake has a discontinuous range; one population occurs in southern South Carolina and into eastern Georgia, whereas the other population ranges throughout Florida into southern Georgia. The species does not occur along the coast from the Georgia-South Carolina border to the northeastern corner of Florida.

Island distribution: The Florida green watersnake is not known to occur presently on our barrier islands.

Preferred habitat type(s): On the mainland, freshwater wetlands of all descriptions support populations of this snake. Open sloughs, ponds, lakes, and canals, often those with marginal emergent and floating vegetation that provide basking sites, are favored by this species in Southwest Florida.

Size: *Nerodia floridana* is the largest North American watersnake. Adults average 76 to 140 cm (30 to 55 in); the record length for this species is 188 cm (74 in).

Color: At birth, the neonate Florida green watersnake has a base color of drab green with intermittent darker dorsal spotting. At maturity, the adult is uniformly pigmented with a dark green hue or an olive drab, almost dull military-like coloration. Its venter is typically white or a pale olive. Some specimens may have small spots on the edges of the wide ventral scales.

Other characteristics: The large Florida green watersnake is very heavy bodied. Like the other watersnakes, it is predisposed to resent handling and can inflict painful but non-venomous bites.

Diet: Fish and amphibians are the primary foods of this species. It is attracted to the odor of periodic fish kills, where it congregates in large numbers and consumes dead fish.

Reproduction: The Florida green watersnake is ovoviviparous and known for its large brood sizes. There is a brood-size of 101 young on record produced by an exceptionally large *Nerodia floridana*.

Life history: As with other *Nerodia*, this snake is closely associated with its aquatic environment and seldom ventures far from water. Its daily life is frequently disrupted by flooding in wetlands, after which it becomes more prone toward movements and thus exposed to mortality on roads and highways. This snake is regularly nocturnal and spends most of its daylight hours basking on emergent aquatic vegetation or on low shrubs that are very close to the edge of the water. When threatened, this snake dives into the water and conceals itself remarkably well, often hiding unseen in surface-floating plants.

Range Map 19. Collection site of the lone specimen of *Nerodia floridana* from Sanibel Island in 1959.

Population status: In wetlands that are some distance from roadways, this species is relatively safe. Some wetlands, however, have been destroyed after alteration by massive drainage projects. With reduction in its preferred habitat, the Florida green watersnake continues to decline.

Threats: The threats to all watersnakes are similar. Drainage and alteration of habitat, removal of cover on the banks of subdivision wetlands and retention ponds that have replaced natural bodies of water, fragmentation of wetland habitat, and a subsequent increase in the number of road kills continue to impact these snakes negatively.

Comments: LeBuff collected a live adult *Nerodia floridana* upland of the water hyacinth wrack line on Point Ybel beach, the easternmost projection of Sanibel Island, in 1959 (LeBuff 1959). *Nerodia floridana* has not been observed or collected on Sanibel Island since.

Nerodia taxispilota (Holbrook, 1838)
Brown watersnake

The brown watersnake seldom ventures far from fresh water, and this species usually selects basking sites where it can slip into the water whenever it feels threatened by an approaching intruder. Since it often climbs and reaches branches overhanging water, this species sometimes drops into passing boats. Because of its broad bands and coloration, the brown watersnake is often misidentified as a cottonmouth. *Bill Love*

Other common names: Water moccasin, moccasin.

Similar species: Watersnakes that share habitat with this species are relatively similar in Florida, as is the venomous cottonmouth. All are different in coloration, however. The cottonmouth differs from the harmless watersnakes in more than color and markings. The eyes of the cottonmouth have vertical pupils; those of the harmless species are round. The pronounced facial pit, which identifies it as a pit viper, is situated on each side of a cottonmouth's head between the eye and nostril. Harmless snakes lack these sensory pits.

General range of species: The brown watersnake is confined to the southeastern states and ranges across the region diagonally from southeastern Virginia to southern Alabama. It occurs throughout Florida.

Island distribution: This species is not known to occur presently on our barrier islands.

Preferred habitat type(s): *Nerodia taxispilota* is a highly aquatic species that occurs along slow-moving rivers and large creeks, in cypress-dominated systems with permanent central ponds, along the margins of large lakes, through expansive sawgrass prairies that have scattered open ponds, and in extensive and wide roadside borrow canals.

Size: The brown watersnake is a large watersnake, with adults averaging 76 to 152 cm (30 to 60 in). The record total length for this species is 176.6 cm (69.5 in). The adult female usually attains a greater total length than the male.

Color: *Nerodia taxispilota* has a basic ground color of light tan on which a dorsal pattern of square-shaped darker brown blotches are superimposed. Blotches of similar dark brown are spaced along the lateral sides of the snake's body. These rise from the light-colored base that has dark blotches typical of this species' ventral (bottom) surface.

Other characteristics: When caught or cornered, the brown watersnake usually has a pugnacious attitude. For a nonvenomous species, it is capable of inflicting a painful bite. This species is an exceptionally good climber, and individuals frequently bask on tree limbs overhanging water. When frightened, it drops into the water, sometimes falling into passing boats where its aggressive behavior and slight resemblance to the cottonmouth have earned the species a bad reputation among surprised boaters and fishermen.

Diet: Fish and amphibians are the chief prey of the brown watersnake.

Reproduction: *Nerodia taxispilota* is an ovoviviparous species like other natricine snakes (which include the watersnakes) and females produce huge broods of young. Broods can total as many as 60 neonates at a single birthing.

Life history: The brown watersnake is more diurnal than most others of this genus. In some habitats, it is abundant, whereas in other similar habitats it is much less common. When not foraging, much of this snake's time is spent basking on elevated sites over or close to water into which they seek refuge if frightened or approached. It has been recorded to climb frequently as high as 6.1 m (20 ft) into the branches of tall trees overhanging water.

When handled, the brown watersnake discharges quantities of fecal matter from its vent and simultaneously releases powerful, foul-smelling, and long-lasting musk from glands in its anal region. Because of its resemblance to the venomous cottonmouth, many of these snakes and their semi-aquatic relatives are killed by people who are unable, or refuse to learn how, to distinguish the different snakes.

Population status: Whereas this snake's abundance has been historically robust, its numbers are unknown. This snake's density continues to decline, especially in habitats that have been surrounded by ever-encroaching suburbia.

Threats: Habitat destruction because of land development and habitat splintering by road networks and associated road kills continue to impact populations of brown watersnakes.

Comments: *Nerodia taxispilota* has never been common in Southwest Florida. LeBuff observed and collected only two specimens during his collecting career. One of these was taken in a borrow pond near Pinecrest in Monroe County (1956), and the other from the Tamiami Canal, near Ochopee, in Collier County (1957). For more than four years (1954-1959) LeBuff and others of the Collier County Herpetological Society regularly monitored live, injured, and DOR amphibians and reptiles on the pavement of the Tamiami Trail (U.S. 41) between Naples, in Collier County, and Forty-Mile Bend in Dade County whenever environmental conditions were conducive to successful night collecting. Animals were collected on the pavement or shoulder after being driven out of the flooded wetlands on both sides of the highway as they were trying to reach higher ground. Thousands of natricine snakes (watersnakes and gartersnakes) were collected alive or DOR on this highway. An individual's location, its identification, and its size was recorded. Neither LeBuff nor any of his associates ever collected a brown watersnake from the shoulder or surface of this highway. The diurnal habits of this snake could account for the paucity of nighttime observations, but during wetland flooding individuals of all other species known to use this habitat were found. Neither LeBuff nor Lechowicz has ever observed *Nerodia taxispilota* on Sanibel Island.

    USFWS Refuge Officer Lennie Jones, now retired, added this watersnake to the JNDDNWR Amphibian and Reptile List for Sanibel Island, but the year of collection and pertinent locality data are unavailable. After LeBuff inquired as to specifics about the Sanibel Island documentation in 2009, Jones advised (pers. comm.) that he was completely familiar with the species and had collected and recorded the brown watersnake on Sanibel Island but had forgotten the specific locality data. A search through JNDDNWR files in 2009 failed to provide further details. As with the collection of the green watersnake at the SILS by LeBuff in 1959, Jones' documentation of the brown watersnake on Sanibel Island may have been based on the sighting of a lone individual.

Pantherophis alleghaniensis (Holbrook, 1836)
Eastern ratsnake

A small adult eastern (yellow) ratsnake at rest in a group of rosary peas (*Abrus precatorius*). These colorful seeds, now common in Florida, are produced by this exotic, well-established perennial vine from India. The snake is harmless, but the seeds are deadly poisonous to all animal species if chewed or swallowed. *Bill Love*

Other common names: Yellow ratsnake, striped chicken snake, chicken snake.

Similar species: Hatchlings are similar in pattern to those of the cornsnake, and in some respects resemble neonates of the black racer and coachwhip. Light-colored adults could be confused with the coachwhip.

General range of species: The yellow-color phase of the eastern ratsnake is confined to the southeastern U.S. This snake's distribution includes that part of the country that is situated along the coastal plain between southern North Carolina and the northeastern section of north Florida. In the southern section, its range broadly expands to include the entire lower two-thirds of the Florida peninsula.

Island distribution: *Pantherophis alleghaniensis occurs* islandwide and is found in all ecosystems.

Preferred habitat type(s): This species is typically found in abandoned buildings, oak- and cabbage palm-forested ridges, and along the margins of wetlands. The adult ratsnake regularly establishes itself within colonial bird rookeries where it preys on bird eggs and nestlings.

Size: This is a large snake with an adult size that ranges from 107 to 183 cm (42 to 72 in). The length record for this snake is 221 cm (87 in).

Color: From the blotched pattern typical of neonates, the yellow ratsnake will attain a variable adult coloration that is basically yellow. As its juvenile blotches begin to disappear, specimens transition in appearance and attain the typical adult pattern that consists of four narrow but well-defined longitudinal brown-hued stripes positioned on the lateral and dorsal surfaces of the body. In most old adults the blotches are completely obscured. The ground color can be variable from golden yellow, to orange, to tan. Some populations in sections of the upper peninsula of Florida have a greenish-toned ground color.

Other characteristics: A powerful constrictor, this snake quickly captures and overcomes its prey. When an individual is encountered in the field by a human and cut off from escape, *Pantherophis alleghaniensis* vigorously defends itself by striking repeatedly at its antagonist. This ratsnake prefers to withdraw from confrontation whenever possible, however. If a large struggling adult ratsnake is picked up by a person, its strength is impressive as it tries to free itself. When a captured specimen wraps around and squeezes a hand or wrist, the exerted pressure is forceful.

A hatchling eastern (yellow) ratsnake caught in the act of consuming a brown anole (*Anolis sagrei*). Snakes have the ability to dislocate their jaws to allow oversized prey to be eaten with ease. Through time, the blotches on this neonate will transform into four longitudinal stripes. *Daniel Parker*

Diet: Juveniles feed on invertebrates and lizards. Through time, the ratsnake graduates to small mammals, bird eggs, and nestling birds. One of its common names is based on its habit of invading chicken coops and preying on the eggs and young of the occupants.

Reproduction: *Pantherophis alleghaniensis* is oviparous and deposits from five to 27 eggs in late summer. Eggs are elongated, measuring on average 18.0 by 40.0 mm (0.7 by 1.57 in), and hatch in approximately 65 days.

Life history: The yellow ratsnake is regularly arboreal and spends much of its time in trees and inside cavities or other secure locations in large trees. It also frequents old buildings or accessory structures, where it remains partially hidden in niches in the structural framing.

Population status: This ratsnake continues to do well on Sanibel and Captiva islands and is benefiting because of habitat protection policies on the islands. The eastern ratsnake was considered the more common of the two ratsnakes on Sanibel Island for the past half century. It now appears that the cornsnake may outnumber the eastern ratsnake in many areas of Sanibel Island, especially on the eastern end of the island.

Threats: Fragmentation of habitat impacts juveniles of this species when they are exposed in residential yards or while crossing roads.

Comments: Commonly known on the islands as the yellow ratsnake, *Pantherophis alleghaniensis* was well established on Sanibel Island and locally common in 1959. Among this species' favorite habitats were the old farmhouses and packinghouses that were still standing on the island in those days, remnants of early 20th century farming operations. The ratsnake remains relatively common on the island, and because the adult usually remains in its established territory, it is less prone to deadly vehicular traffic.

Pantherophis guttatus (Linnaeus, 1766)
Red cornsnake

In Southwest Florida, the red cornsnake (*Pantherophis guttatus*) is generally known as the red ratsnake. This native species is considered a relatively recent newcomer to the barrier islands and is now established throughout Sanibel Island. *Daniel Parker*

Other common names: Corn snake, red ratsnake.

Similar species: The coralsnake, scarlet kingsnake, and scarlet snakes also have a red-toned coloration. The cornsnake is very similar in appearance to other ratsnakes, at least for the first year of its life.

General range of species: The cornsnake ranges across the breadth of the lower and upper coastal plains of the southeastern states from southern New Jersey to Mississippi and southeastern Louisiana. This species occurs throughout the Florida peninsula.

Island distribution: *Pantherophis guttatus* is established islandwide on Sanibel Island and has been observed or collected in all ecosystems.

Preferred habitat type(s): The cornsnake resides in a variety of habitats. In some parts of its range, it frequents corn cribs where it preys on rodents, hence its common name. On the barrier islands it is frequently found near buildings and in West Indian/cabbage-palm hammocks.

Size: The red cornsnake is a moderate sized snake that grows from a 305-mm (12-in) hatchling to an average adult size of 122 cm (48 in). The maximum recorded length for this species is 183 cm (72 in).

Color: The color characteristics of *Pantherophis guttatus* are highly variable, and some populations in localized habitats are colored much different from the standard for this species. The cornsnake on the islands is a descendent of that on the adjacent mainland. It typically has a silver-gray ground color with a series of red-hued dorsal and lateral blotches. These blotches are boldly outlined with a thin black line. Its bottom ventral surface has a vague black and white pattern, almost like a checkerboard.

Other characteristics: The cornsnake spends more time on the ground than most of the other ratsnakes, and this snake often follows tunnels and explores rodent burrows in search of prey. It is an excellent climber, and on rough-barked trees can virtually climb up the trunk in a straight line without using typical serpentine undulations.

Diet: Small mammals, nestling birds, and bird eggs constitute the major prey of this snake.

Reproduction: The cornsnake is an egg layer. The female deposits 10 to 30 eggs in late spring or early summer. Eggs are elongated and on average measure 19 by 38.1 mm (0.75 by 1.5 in). They hatch in 60 to 65 days.

A hatchling red cornsnake. This individual will retain its basic pattern into adulthood, but the overall coloration will change. *Daniel Parker*

Life history: *Pantherophis guttatus* is similar in habits to its relative, the eastern ratsnake. The cornsnake is skillfully arboreal and will climb into the rafters of old buildings and the canopy of trees to reach prey.

Population status: This snake's abundance is generally declining throughout its range. Any increase in numbers is negated by the expansion of habitat destruction because of agricultural and residential development.

Threats: As with other large snakes, this species is often killed by people when individuals venture into residential yards. Habitat fragmentation continues to destroy cornsnake populations in both suburban and rural habitats.

Comments: This snake was first reported from the eastern end of Sanibel Island, circa 1964. In the years post-causeway (after 1963), the red cornsnake slowly became common on the eastern half of Sanibel Island by 1975. By 2012, it had dispersed islandwide in suitable habitat. The cornsnake has always occurred in moderate numbers on the nearby mainland around Iona.

In 1990, Sanibel native Ralph Woodring showed LeBuff the frozen head and neck of an adult cornsnake that one of his employees had killed. Woodring had never observed this species prior to finding this individual, and he had roamed all ecosystems in his 55 years of island living at that time.

Lechowicz collected a gravid specimen of *Pantherophis guttatus* from SCCF's Frannie's Preserve on Sanibel Island in 2004 and allowed it to deposit eggs in captivity. One of the hatchlings was anerythristic (lacking any red pigment). This is a very popular color morph in the pet trade. Most of the cornsnakes sold by captive breeders have multiple genetic recessive traits that show up in future offspring if they are mated with another heterozygous wild type. The production of the anerythristic hatchling suggests, as first hypothesized by LeBuff, that cornsnakes were introduced to the island either as stowaways on shipments or on purpose as released pets. The type locality for the anerythristic characteristic is near Immokalee, Collier County, Florida.

Ramphotyphlops braminus (Daudin, 1803)
Brahminy blindsnake (Passively introduced)

The tiny, nonnative, brahminy blindsnake. It is seldom seen on the ground surface, but this species is now very common on the barrier islands. It is widely distributed throughout the upland habitats provided by landscaped subdivisions. *Bill Love*

Other common names: Worm snake.

Similar species: None.

General range of species: The brahminy blindsnake is an exotic species from Africa and Asia. It has been transported around the world in potted plants.

Island distribution: This snake is found throughout Sanibel Island.

Preferred habitat type(s): *Ramphotyphlops braminus* occurs in loose, moist, shady substrates. Specimens are often encountered in potted plants and debris piles in and around residences. This snake is completely fossorial and rarely comes to the surface except at night. It is found in pantropical climates.

Size: This is a tiny snake, averaging 12 to 17 cm (4.7 to 6.7 in).

Color: This blindsnake is dark brown to black on the dorsal side and slightly lighter in color on the ventral side. Just before it molts, the body color takes on a temporary silver-gray hue.

Other characteristics: *Ramphotyphlops braminus* has a blunt head and short tail; it appears very much earthworm-like. It has very small eyes that are often hard to see without using magnification.

Diet: The brahminy blindsnake eats mostly small larval ants, termites, and other soft-bodied insects.

Reproduction: This snake is believed to be an all-female species (parthenogenetic). It can deposit one to eight eggs in a clutch, and the hatchlings are identical to the adult.

Life history: The brahminy blindsnake is rarely seen above the surface of the ground. It is subject to desiccation very quickly, so its activity at the surface occurs during rainstorms and at night. It is completely fossorial and can create a burrow and disappear very quickly.

Population status: This snake can be quite common in residential areas and plant nurseries on the islands.

Threats: This is a nonnative species in the U.S., so threats to its existence have not been considered. There is no evidence that this species is competing with any similar, native species for food or habitat. It is eaten by a variety of reptiles, birds, and mammals. Although there is no supporting evidence, this exotic is likely prey of the southern ring-necked snake and coralsnake on our islands.

Range Map 20. Collection site of *Ramphotyphlops braminus* in 1982.

Comments: This tiny asexual snake of Asian origin first appeared on Sanibel Island circa 1982 (S. Phillips, pers. comm.). Steve Phillips collected a specimen of *Ramphotyphlops braminus* from beneath a group of landscape timbers in the Gumbo Limbo subdivision. By 2000, the range of this population had expanded throughout the neighborhood and at least 0.4 km (0.25 mi) to the east from Phillips' collection site. Other introductions likely occurred at about this same time but went undetected prior to 1982. In 2012, *Ramphotyphlops braminus* continues to thrive on Sanibel Island, an example of another successful passive introduction via potting soil and landscape plants that have been transported across the Sanibel Causeway.

Storeria victa Hay, 1892
Florida brownsnake

The Florida brownsnake is highly secretive and spends most of its life concealed in leaf litter in upland habitats and their moist margins. Although populations of this snake may be common, individuals are usually encountered only during landscaping or torrential rains when the snakes are flooded out of their hiding places. *Daniel Parker*

Other common names: Brown snake, DeKay's snake.

Similar species: Southern ringneck snake.

General range of species: *Storeria victa* ranges from extreme southeastern Georgia south to include all of the Florida peninsula.

Island distribution: This snake is found sparsely throughout Sanibel Island in dry and moist habitats in or near woodlands.

Preferred habitat type(s): The preferred habitat of *Storeria victa* is a wooded area with ample debris for hiding. This snake can be found alongside water sources or in upland areas. It is almost always associated with nearby woodlands but is also found in vegetated residential areas and is fond of moving around under the surface of loose landscape mulch.

Size: The Florida brownsnake is a small species, 22 to 33 cm (9 to 13 in), with a maximum recorded size of 48.3 cm (19 in).

Color: The brownsnake gets its name from its coloring, which is mostly some shade of brown, although it can be gray or even red-toned. The head is usually a darker shade of brown than the body. The neck band is creamy white or yellow-hued.

Other characteristics: *Storeria victa* has two rows of black spots on the venter.

Diet: The Florida brownsnake eats mostly earthworms and slugs but has been documented eating fish, amphibian larvae, and insects.

Reproduction: *Storeria victa* is ovoviviparous, meaning that its eggs incubate internally and the young are born alive instead of hatched. A litter can have from five to 40 young.

Life history: This brownsnake is common in many parts of Florida. *Storeria victa* is a snake of woodland areas and is often encountered while lifting debris or leaf litter. Its only defense is the musk that it produces when frightened.

Population status: The Florida brownsnake is not very common on Sanibel and Captiva islands. It has been documented at least once each year, but not many are found.

Threats: This species is eaten by mammals, birds, and other snakes. It has abundant habitat on the islands. Its secretive nature may explain why it is not commonly recorded.

Comments: *Storeria victa* was not included on the original 1959 amphibian and reptile list for SNWR. The species' presence was not documented on Sanibel Island until 1960 when a live specimen was collected as it was crossing Periwinkle Way during torrential rainfall. This snake continues to be an uncommon and secretive species in the island's Mid-Island Ridge Zone and similar habitats.

Thamnophis sauritus sackenii (Kennicott, 1859)
Peninsula ribbonsnake

The peninsula ribbonsnake has always been common in the interior wetlands of Sanibel Island. The dorsal stripe in this subspecies is usually obscure, as is that of the specimen pictured; in some individuals, it may be entirely absent. *Daniel Parker*

Other common names: Ribbon snake, garter snake.

Similar species: Eastern gartersnake.

General range of species: *Thamnophis sauritus* is found from isolated populations in Wisconsin to Maine in the north and from Louisiana to the Florida Keys in the south.

Island distribution: This ribbonsnake is found throughout the islands, usually on land but near freshwater basins.

Preferred habitat type(s): The peninsula ribbonsnake prefers wet areas with much low vegetation. It readily climbs into bushes but is usually not far from water.

Size: This is a medium-sized snake, 40 to 90 cm (15.7 to 35.4 in). The record length for this species is 101.6 cm (40 in).

Color: The peninsula ribbonsnake is mostly brown to almost black with three yellow-hued stripes on the dorsal side (one down the middle and one on each side of the body). In this subspecies, the central dorsal stripe is sometimes very faint or even completely obscure.

Other characteristics: The stripes on each side of the body can be blue-toned in some areas in Florida. The ventral side is usually yellow or greenish.

Diet: The peninsula ribbonsnake prefers to eat amphibians on Sanibel and Captiva islands. It also occasionally eats insects.

Reproduction: *Thamnophis sauritus sackenii* is ovoviviparous, meaning that the eggs incubate internally and the young are born alive instead of hatching. The female produces four to 25 young per litter.

Life history: The ribbonsnake is common in many parts of Florida. It is usually diurnal and is often seen basking, though a preference for feeding on frogs may shift its daily activity pattern nocturnally so it can prey on breeding frogs.

Population status: The peninsula ribbonsnake is common on Sanibel and Captiva islands. It is seen almost daily on the Sanibel-Captiva Road bike path.

Threats: This species is eaten by mammals, birds, and other snakes. It has abundant habitat on the islands.

Comments: In 1959, *Thamnophis sauritus sackenii* was regularly observed both alive and dead on the few roads that spanned the Sanibel Slough. Today, this species continues to occur in interior freshwater wetlands, but the frequency of observations is noticeably reduced. Many people, especially from northern regions, confuse it with the gartersnake. The ribbonsnake and gartersnake are closely related.

Thamnophis sirtalis sirtalis (Linnaeus, 1758)
Eastern gartersnake

The eastern gartersnake is one of the most common snakes in the eastern U.S. This species' occurrence on Sanibel Island is known from only two specimens, and it is now considered extirpated on the island. The reason that the peninsula ribbonsnake has been so successful on the barrier islands, while this related snake has not is unknown. *Bill Love*

Other common names: Common garter snake, garden snake, ribbon snake.

Similar species: Peninsula ribbonsnake.

General range of species: This gartersnake is found throughout the U.S. with the exception of the Great Basin and the deserts of the Southwest. A disjunct population of this species occurs in New Mexico. It is also found in southern Canada from coast to coast. The subspecies once found on Sanibel and Captiva islands, *Thamnophis s. sirtalis*, occurs along the east coast of the U.S., throughout Florida, and as far west as east Texas and extreme southeast Saskatchewan, Canada.

Island distribution: The eastern gartersnake is extirpated on Sanibel and Captiva islands.

Preferred habitat type(s): The preferred habitat type of *Thamnophis sirtalis* is in vegetation near a water source. It prefers moist areas along the edge of forests and tall vegetation where it often hides under debris such as fallen logs, rock piles, or garbage. It does well in suburban neighborhoods.

Size: The eastern gartersnake is a medium-sized snake that averages 15.2 cm (6 in) at birth, with adults averaging 50 cm (20 in) in total length. The maximum recorded size of this gartersnake is 123.8 cm (48.75 in).

Color: The eastern gartersnake can be black, gray, or green with yellow or white longitudinal stripes. The head is often slightly lighter in color than the body. The ventral side is yellow- or cream-colored. There are scattered small populations of this snake that lack stripes and are totally black; these are termed melanistic.

Other characteristics: Some eastern gartersnakes have longitudinal stripes of reduced width. This occurs in some of the gartersnakes in South Florida; and some individuals have prominent dark blotches between the stripes.

Diet: The gartersnake eats worms, slugs, fish, insects, amphibians, and occasionally, small birds and mammals.

Reproduction: *Thamnophis sirtalis* is viviparous, meaning it does not deposit eggs. It can have multiple litters of approximately 27 neonates in a season; the record litter for an individual female is 85 (Martof 1954).

Life history: *Thamnophis sirtalis* is a common snake in many parts of the country. It is a diurnal species well known for its cold tolerance in the north. This snake is able to freeze for a limited amount of time and will survive if thawed. This is not an important adaptation in South Florida, but because it is able to function and eat at very low temperatures, it is active most of the year. When provoked and its escape routes are blocked, this species is capable of flattening its body and striking repeatedly.

Population status: The eastern gartersnake was found on Sanibel Island only twice between 1960 and 1978. It has not been documented since then.

Threats: This species is eaten by mammals, birds, and other snakes. It would probably be a successful species on Sanibel and Captiva islands if it were to colonize the islands in numbers.

Comments: *Thamnophis sirtalis* was first documented on Sanibel Island when a DOR specimen was discovered on the Tarpon Bay Road boundary of the SNWR BT in 1960. Another dead eastern gartersnake later was collected in the interior wetlands (Campbell 1978). Apparently never common, the eastern gartersnake has not been documented on Sanibel Island for more than 30 years (Lechowicz 2009).

~SUPPLEMENTAL SPECIES LIST~

The following is a supplemental list of Florida native and nonnative snakes that are non-indigenous to the barrier islands but that have been reported on either Sanibel or Captiva Island since 1959. None of these has become successfully established, and some reports are based on vague observations or individual specimens that we consider to have been released and were observed only once.

~ORDER SQUAMATA—THE SNAKES~

Agkistrodon piscivorus conanti Gloyd, 1969
Florida cottonmouth (Dubious reports)

An adult Florida cottonmouth coiled defensively with its mouth agape. The white interior of this snake's mouth has earned it its common name. The wide brown lateral stripe through its eye is one of the keys to telling the difference between this and the harmless natricine watersnakes. The cottonmouth's sheath-covered, folded-back, retractile fangs are visible inside its mouth on each side of the upper jaw.
Daniel Parker

Also known as cottonmouth moccasin and water moccasin, this snake and the more northerly distributed eastern cottonmouth are almost identical. Any of the watersnakes of the genus *Nerodia* appear very similar to this species when viewed by inexperienced observers. The Florida cottonmouth is distributed from extreme southern Georgia throughout the Florida peninsula.

This snake favors freshwater wetlands including the margins of lakes, ponds, sloughs, cypress systems, and canals. Some localized populations of this species occur in, or very close to, brackish water habitats.

Agkistrodon piscivorous conanti is a large pit viper, with adults averaging 76 to 122 cm (30 to 48 in) in length. The record length for this subspecies is 189.2 cm (74.5 in). The largest cottonmouth ever caught by LeBuff (1958) came from the Tamiami canal in Collier County and measured 162.6 cm (64 in) in total length.

The facial pit of the cottonmouth, located between the eye and nostril, identifies this snake as a pit viper and further distinguishes it from the nonvenomous watersnakes. This snake possesses venom that is basically hemotoxic. It destroys blood cells and tissue and hinders the ability of the victim's blood to coagulate. Cottonmouth bites have qualities that often result in a serious gangrenous outcome at the site of envenomation. This snake's venom is slightly different from that of the rattlesnakes, which also have hemotoxic venom. By volume, rattlesnake venom has more neurotoxic components than does that of the cottonmouth.

This snake usually reacts to a threat by quickly coiling and positioning its head near the center of the coil. It then opens its mouth widely to exhibit a fully white interior, perhaps warning intruders to keep their distance while at optimal position to inflict a serious bite.

Neonate Florida cottonmouths are vividly colored in contrasting earth tones. The identifying dark head stripe stands out as does the yellow tail of this young snake. The tip of its tail is often used as a lure to attract prey. This color fades away as the snake ages as do the visible crossbands. *Bill Love*

The cottonmouth remains associated with wet habitats throughout its life, although specimens are sometimes encountered at a considerable distance from water. This snake is sometimes active during daylight but is generally nocturnal, spending the daytime basking near the bank of its wetland habitat, on tussocks within those systems, or completely concealed. On the mainland we have observed that the cottonmouth becomes active at dusk and seems to materialize instantly, almost out of nowhere, as it leaves its hiding place and begins to forage. This occurs especially in the spring during periods of low water in mainland cypress habitats where the cottonmouth is abundant.

Carr (1940) mentioned that the Florida cottonmouth occurs on some of the "high" islands along Florida's gulf coast. More recently, Lillywhite and McCleary (2008) affirmed that this species inhabits some of the islands along the coast. In particular, they reference the large numbers of this subspecies that occur on Sea Horse Key in Levy County, Florida.

Over the past 55 years, there have been only two reported sightings (undocumented) of cottonmouths anywhere on Sanibel Island (W. Hammond, pers. comm.; R. Loflin, pers. comm.)[37]. Some large, very dark specimens of the Florida watersnake (*Nerodia fasciata pictiventris*) inhabit the areas where these observations were reported. There are no documented cottonmouth reports from any of the major islands of Pine Island Sound to this day (K. Krysko, pers, comm.).

From 1959 through 1991, LeBuff spent thousands of hours in the field both day and night and frequently sampled all habitats at night by canoe or motor vehicle without ever finding a cottonmouth on Sanibel. Neither he nor his colleagues were ever concerned about venomous snakes while they blindly tromped through the Sanibel Island wetlands by day or night. This was counter to his perception when afield in the Big Cypress Basin on the mainland, where every step was measured with a scanned evaluation of anticipated footfalls, and elevated vegetation in his path was carefully examined before passing.

This venomous snake may indeed be collected and documented on Sanibel Island someday. It is difficult to understand why its nonvenomous counterpart, *Nerodia fasciata pictiventris*, has successfully colonized the island's freshwater wetlands while the cottonmouth has not.

Sistrurus miliarius barbouri Gloyd, 1935

Dusky pigmy rattlesnake (Dubious report)

Most venomous snakebites in Florida occur when people come into contact with the dusky pigmy rattlesnake. In recent years, many bites have resulted when these snakes were encountered in the plant section of large discount chain stores. These small rattlesnakes are transported in containers along with field-grown nursery stock. The venom of this snake is extremely toxic. The rattle of the adult pictured above is a tiny structure that is not loud enough to be heard more than a few meters away. *Bill Love*

Sistrurus miliarius barbouri is a snake of the southeast occurring from southern Georgia to the Florida panhandle and through the entire peninsula. It is commonly called pigmy rattlesnake, ground rattlesnake, and ground rattler. As the name "pigmy" implies, all members of the genus *Sistrurus* are relatively small rattlesnakes. Adults average 50 cm (19.7 in), and the record length for the pigmy rattlesnake is 80.3 cm (31.5 in).

The dusky pigmy rattlesnake generally has a gray ground color over which darker gray blotches, each outlined in a lighter gray, are present dorsally. A series of three rows of gray spots adorn the lateral surfaces. A narrow red-brown dorsal stripe extends from the neck to the tail.

This rattlesnake is responsible for many of the venomous snakebites that occur in Florida each year. There are no documented cases of human fatalities from its bite; however, its bite is extremely painful and is accompanied by severe tissue damage. Victims of *Sistrurus* envenomation on their hands may end up with fingers that are crooked and misshapen 60 years after the event (R. Curtis, pers. comm.).

The pigmy rattlesnake is threatened wherever it occurs. It is usually killed on sight by people who can't shake the belief that "the only good snake is a dead snake."

A voucher specimen of the dusky pigmy rattlesnake has yet to be collected or photographed for documentation of this species' occurrence on Sanibel Island. In the mid-1980s, a member of the COS's Wildlife Committee announced at a meeting that two dusky pigmy rattlesnakes had been found at Casa Ybel Resort on Sanibel Island. This report was investigated and negated when the snakes turned out to be misidentified hatchling black racers, *Coluber constrictor priapus*.

Sistrurus miliarius barbouri is commonly transported in potted landscape trees and plants throughout Florida. Individuals are frequently discovered in potted plant containers. It could be only a matter of time before this species is discovered on Sanibel Island.

Boa constrictor Linnaeus, 1758

Boa constrictor (Escapee)

Many species of snakes have developed complex and high levels of protective coloration over time. The camouflage pattern on the adult *Boa constrictor* is a graphic example of how a snake's pattern serves its survival. As with other giant constrictors, this species is becoming more frequently encountered in Florida's wild lands. *Bill Love*

The boa constrictor ranges from northern Mexico to Argentina and occurs on many islands in Central and South America. This snake is considered one of the giant constrictors, which include the Burmese python.

Boa constrictor occurs in a variety of habitats, from tropical rain forests to the fringes of deserts. In its native habitat it frequents wetlands, woodlands, and savannahs. Adults average 2.7 m (9 ft); the record length for this species of 5.64 m (18.5 ft) has been brought into question and is today considered invalid. Depending on an individual's size, this boa will consume small- to medium-size prey, including lizards, birds, rodents, and monkeys. Large specimens of this snake are capable of eating small feral swine.

The boa constrictor is an attractive snake, colored with shades of tans, browns, greens, and yellows—even bright reds. One subspecies has an attractive light tan ground color with sharply distinctive dorsal lines and saddle-like blotches that become nearly bright red near the posterior of the snake's body. Thus, one of this species' common names is red-tailed boa[38].

The sheer size of *Boa constrictor* and the few other giant members of the snake families Boidae and Pythonidae, which are increasing in numbers in Florida, make their identification easy as an undesirable exotic species.

In 1978 SCCF Director of Environmental Programs Steve Phillips (pers. comm.) responded to a call from a resident concerning the sighting of a large snake on western Sanibel Island in the area of Blind Pass Condominium (Bowman's Beach Road). When he arrived on the scene he discovered and captured an adult *Boa constrictor* that approached 3 m (9.8 ft) in length. After researching the possible origins of this snake, Phillips learned that a 1.8-m- (6-ft)-long boa constrictor had escaped in 1974 from a private reptile collection near the junction of Sanibel-Captiva Road and Bowman's Beach Road. Sanibel Island's habitat had accommodated this specimen very well.

Boa constrictor has already become naturalized in southeastern Florida and has joined the introduced pythons[39] as they continue to invade the state's natural systems where threats to control its invasion are usually absent. This boa has been established for more than 20 years on the Deering Estate in Miami, Dade County, Florida.

Python molurus bivittatus (Kuhl, 1820)

Burmese python (Releasee)

A 2.4-m (8-ft) Burmese python that was most likely conceived and hatched in the Florida Everglades. Larger adults, up to 5.69 m (18 ft 8 in[40]), are regularly encountered and collected or killed in South Florida. Eventually, this snake will seriously impact the native wildlife of the Florida peninsula, and some of the animals on which this python preys are currently seriously endangered. The FFWCC and other agencies must attack this problem realistically. We suggest that this body becomes proactive and open the python-control door of opportunity wider by adopting a new policy that provides for the general and unrestricted taking of this snake in Florida. This should be coupled with no-fee-required training seminars and registration for would-be python exterminators. To be successful, this would work best as a bounty-based program in which any permanent Florida resident or landowner can properly enroll in the program and participate in the war against pythons without any licensing fees. *Bill Love*

The native distribution of the Burmese python spans the Indian subcontinent. This snake ranges from eastern India, through northern mainland Malaysia, southern China, and into Indonesia. This python is one of the giant snakes, and considered to be the sixth-largest snake species in the world. At hatching it averages 50.8 cm (20 in) and grows rapidly. Adults average 3.7 m (12 ft), and the species is known to reach a record length of 5.8 m (19 ft).

Python molurus bivittatus swims well and is usually found close to water; thus, it does well in the Florida Everglades and adjacent wetland systems. Other than enormity of size, other unique features of this snake are its giraffe-like markings and the presence of a dark arrowhead-like marking atop the head. The latter feature points toward and extends nearly halfway to the tip of the animal's snout. The Burmese python preys on a variety of mammals, birds, and reptiles. Although mammals appear to be the prey of choice, this snake consumes basically any organism it can swallow. There have been no documented cases that include humans in this species' prey base, although, rarely, captive Burmese pythons have caused human mortality. In Florida, this snake's prey has been verified to include adult deer and alligators.

The naturalized population of Burmese pythons in southern Florida may be steadily increasing; however, many herpetologists believe its total numbers were set back as a result of low-temperature-induced mortalities during the cold winter of 2009-2010.

In about 1988, an excited visitor entered the JNDDNWR Visitor Center and announced she had just encountered a large Indian (Burmese) python, *Python molurus bivittatus*, and had stopped her vehicle and watched the snake cross the refuge entrance road. She was insistent that her identification was correct because her son kept one of these snakes in captivity. LeBuff and another refuge staff member responded and together they scoured the area where the python had been sighted for an hour but to no avail. In our opinion this was a credible report, and this specimen and possibly other actively released pythons remain undetected on Sanibel Island.

George Campbell (Campbell 1985) collected a large Burmese python on Captiva Island in 1985. The origins of that specimen were never determined.

In 2006-2007 there were reports of a large python or boa at the Gavin Site on JNDDNWR. At that time, the location was being used as a staging area for a refuge project, and it housed stockpiles of gravel and fill. A sprinkler system kept the piled materials moist. The island was dry at the time, and there was little standing water around. The sprinklers formed pools of water that attracted animals when no one was around. Contractors reported several times that they saw an eight- to 12-foot snake drinking water from the pool. At least one refuge employee claimed to have seen the snake. One night at 2200 hours in 2006, SCCF intern Cara Faillace saw the snake crossing Sanibel-Captiva Road directly across from the Gavin Site.

The State of Florida made headway in general control of pythons in South Florida when FFWCC launched the 2013 Python Challenge late in 2012. Payment of a $25.00 registration fee and simple online training authorized participation in the program. Python Challenge was designed to permit public harvest of pythons in four South Florida Wildlife Management Areas. Legal take and euthanasia of pythons was authorized by this permit between 12 January and 10 February 2013. A week before Christmas, 2012, less than 300 permits had been issued, even though LeBuff's personal permit number issued 17 December was #881. FFWCC developed monetary prize categories for both the largest number of pythons collected by any permitee during this "open season," and for the longest python killed during the hunt. In the future, FFWCC should expand this policy and encourage python take by responsible hunters who use the management areas and other state-owned lands in South Florida at any time of year. During the 2013 Python Challenge 68 Burmese pythons were harvested in the general competition. Of these the greatest number taken by an individual permitee was six and the longest snake recorded measured 4.34 m (14 ft 3 in).

The state used the 2013 Python Challenge hunt as a test to determine if a bounty would be beneficial to help in the removal of these invasive reptiles. Many biologists are against a bounty on wild-caught pythons in Florida because they are concerned that unscrupulous individuals (e.g., some commercial collectors) will release more pythons for income should wild pythons become scarce. This does not seem reasonable, unless a very large bounty per snake is enacted. Others have suggested that people may breed the snakes, on the sly, and sell them to the state. This also seems unlikely if it is a small bounty (perhaps $20.00 or less). With the new federal legislation prohibiting the interstate commerce of these large constricting snakes and the ban on importation of these species, their value in the pet trade will undoubtedly dwindle, as will their reproduction rate in captivity. It will no longer be financially sustainable to breed these pythons. Even if someone gives up a large pet Burmese python to the state for a small bounty (while claiming they caught it), isn't it worth it to have one less breeding size python in Florida that could escape or be released because it is not economically feasible to keep it?

~GLOSSARY~

500-year storm event(s): Statistically, a major storm which has a one in 500, or 0.2%, chance of happening in any year.

Abate®: An organic phosphate insecticide effective against the larvae of mosquitoes.

Advertisement sounds: The loud calls that male frogs generate from their vocal sacs to attract females.

Accreting: The natural build-up of a beach due to processes that deposit sand and other solids that are suspended in coastal waters.

Adulticide: A chemical used to kill adult insects, e.g., mosquitoes.

Ad valorem: A tax levied in proportion to the estimated value of a parcel of real property.

Aestivate/aestivation: A period of dormancy, usually during periods of dry and/or hot weather, when some amphibians and reptiles become inactivate and await improved conditions.

Alligator hole(s): An aquatic microhabitat of deep water that is excavated by, and whose size and depth is maintained by, an adult alligator(s).

Alligator Trapper: In Florida, the FFWCC contracts with qualified private individuals who become authorized nuisance alligator trappers and may remove and kill problem alligators.

Amphibian(s): An ecothermic (cold-blooded) vertebrate of the class Amphibia; frogs, toads, newts, and salamanders. These animals usually hatch with an aquatic gill-breathing larval stage that is followed by a lung-breathing adult terrestrial stage.

Anal wart(s): Well-defined raised areas on the skin near the anal opening of certain amphibians.

Anuran(s): The frogs and toads; an amphibian of the Order Anura.

Aquifer: Usually, but not always, referring to other than surface water. Typically a subterranean geological formation which contains voluminous amounts of ground water that supplies the water for wells, springs, etc.

Arboreal: Pertaining to an animal species that dwells in shrubs and trees.

Arribada: A Spanish word which when translated means "arrival;" in herpetological usage, the massive synchronized nesting by sea turtles of the genus *Lepidochelys* and some freshwater turtles of the genus *Podocnemis*.

Artesian well(s): A deep-water well that is drilled through a series of impermeable strata to reach an aquifer of water that, because of internal hydrostatic pressure, is capable of rising many meters to the terrestrial surface or above.

Asexual: To reproduce by parthenogenesis; an unfertilized egg develops into a new individual. Asexual species are always female.

Barrier island: A narrow island, usually positioned parallel to the coast of the mainland, that has accreted over time because of coastal waves, currents, and sediments. These usually long islands function as barriers that in essence protect the mainland from surf- and tidal surge-caused erosion.

Biogenic: As used here, soils produced by the decomposition of calcium-rich organisms, e.g., pulverized seashells and crustaceans.

Biological diversity: The variation of life forms within a specific ecosystem or on the planet in total. Also known as biodiversity; a measure representing the viability of biological systems.

Blindsnake(s): Tiny burrowing insectivorous snakes of the families Leptotyphlopidae and Typhlopidae. These earthworm-like snakes lack a distinct head and have poorly-developed eyes.

Brackish water: A mix of fresh and salt water, usually found in estuaries. Brackish water is water that has more salinity than fresh water, but not as much as seawater; a result from mixing seawater with fresh water (an estuary).

Bufotoxin: A toxic substance that is secreted defensively from the parotoid and other skin glands of toads.

Bycatch: Non-targeted marine creatures that are caught in nets during fishing operations for other species.

C: Abbreviation for Celsius.

CCL: Curved carapace length. A measurement taken over the curve of a turtle's upper shell.

Coastal plain: A sometimes broad plain of low elevation adjacent to a seacoast.

Carapace: The upper shell of a turtle.

Castor bean: The poisonous seed of the castor-oil plant, *Ricinus communis*, from which castor oil is rendered.

Caudal scute(s): The raised keel-like scales along the top of an alligator's tail.

Caudata: The Order of salamanders and newts.

Chelonian(s): Pertaining to the Order Chelonia; turtles.

Chlorpyrifos: A broad-spectrum insecticide used on lawns and ornamental plants.

Chytrid fungus: An often fatal skin disease in frogs caused by the fungus *Batrachochytrium dendrobatidis*. This fungus has been detected on at least 287 species of amphibians from 36 countries, and has caused declines in amphibian populations around the world. It may be responsible for more than 100 of the earth's frog species extinctions since the 1970s.

Circumglobal: To be distributed around the world within a range of latitudes.

Clade: A group of species that are more closely related to each other than any other group, implying a shared common ancestor.

Closure Order: In this sense, a presidential order closing certain lands and waters to hunting; prohibiting the taking of migratory birds.

Cm: Abbreviation for centimeter.

Conchologist: The branch of zoology dealing with the shells of mollusks.

Cooter: Typically in the southern U.S., a freshwater turtle of the genus *Pseudemys*. Believed to be based on the term *kuta*, an African word for turtle from the continent's Bambara and Malinke languages.

Cranial crest(s): A raised ridge on the top of the head of certain toads, located between or behind the eyes.

Crocodilian(s): Large semiaquatic reptiles of the Order Crocodylia; the alligators, caimans, crocodiles, and the gharial.

Crotaline snakes(s): Snakes grouped into the family Crotalidae; the copperhead, cottonmouth, rattlesnakes, and pigmy rattlesnakes.

Curved carapace length: A measurement taken along the dorsal curve of a turtle's upper shell, from between the anterior edge of the nuchal notch to the most posterior extension of the longest rear marginal.

Describer: The individual(s) who describes (names) new genera and species.

Detiologic agent: Any number of disease-causing microscopic organisms such as bacteria or viruses.

Detritus: As used here, refers to any disintegrated vegetative material or debris.

Dewlap: An expandable fold of loose skin in some lizards (especially anoles) which is extended fan-like under their throat and is usually displayed during courtship.

Diazinon: An organophosphorous insecticide.

Diminutive: Extremely small in size.

Disjunct: As used here, refers to an animal's range that is separated into parts.

Diurnal: An animal active in the daytime; also a type of tidal fluctuation.

Documented: A valid record; one based on expert observation, or an identifiable photograph, or a collected specimen from a known locality.

Dorsal side: Refers to the upper surfaces; the back or spinal part of the body.

Dorsolateral ridge(s): Lines or folds of skin (often gold or white) along the upper sides of some frogs.

DOR: A specimen collected Dead On the Road.

Downburst(s): A strong downward current of air flowing from a cumulonimbus cloud and usually accompanying intense rainfall or a thunderstorm.

Drift fence: In biology, a very useful population survey tool consisting of a low fence, about 10-50 meters in length, made from fabric or sheet-metal with collection receptacles usually buried along it and/or at each end. Amphibians and reptiles encountering the fence will turn and follow it, ultimately falling into one of the collection containers or funnel traps from which they cannot escape.

Ecology: The field of biology that studies the relationships between organisms and their environments.

Ecological niche(s): Refers to the role an organism plays within its environment.

Ecosystem(s): An ecosystem is a biotic community considered to be balanced with its physical environment and recognized to be an integrated unit.

Elapid: Venomous snakes of the family Elapidae, which includes the cobras, coralsnakes, kraits, and mambas.

Emigrate: To leave a country or area.

Endangered: A biological organism considered by specialists to be seriously at risk of extinction.

Endangered Species Act (ESA): The Endangered Species Act was passed by the U.S. Congress in 1973 as a federal effort to protect plant and animal species within the states and territories of the U.S. that are at risk of extinction.

Endemic: Species of plants and animals which are found exclusively in a particular area and are naturally not found anywhere else.

Epibiont: An organism that lives on the surface of another living organism. For example, barnacles living on a sea turtle's shell.

Erosion: As used here, the term is associated with beaches and refers to the process by which the surface of the earth is worn away by the action of water and wind.

Ebb tide(s): Refers to what can also be termed as a falling, receding, or outgoing tide.

Endosulfan: A controversial organochlorine compound that is used as an insecticide.

Estuary: A partly enclosed coastal body of water with a river(s) or stream(s) flowing into it, and with a tidal connection to the open sea.

Et al.: An abbreviation indicating that additional persons (authors) have contributed to the cited reference.

Exotic pest plants: Non-native invasive plant species.

Exotic species: An introduced non-indigenous, or non-native species, or an introduction; a species living outside of its native geographical range.

Extant: Refers to a species still in existence; one not extinct.

Extirpated: A localized extinction, where a species no longer exists in a particular area but continues to exist elsewhere within its original range.

Extirpation: Acts that result in extirpating or eradicating; the state of being extirpated or eradicated.

F: Fahrenheit, capitalized when abbreviated.

Family: A taxonomic rank below the order and above the genus in biological classification.

Federal Register: A daily publication of the U.S. government that publishes proposed and final administrative regulations of federal agencies.

Fenthion: An insecticide.

Flatwoods: A terrestrial ecosystem (sometimes referred to as pine flatwoods), the most extensive ecosystem in Florida.

Florida Trustees of the Internal Improvement Trust Fund: The Florida governor and cabinet sit as Trustees and oversee this unique trust as the guardians of sovereign lands in the state.

Fossorial species: An organism that is adapted to digging and spends most of its life underground in burrows.

Form(s): As used here, this is a vague term used in classification when the correct taxonomic rank is not clear.

Frog Watch Network: A volunteer group in Southwest Florida that monitors amphibian populations under protocols of the North American Amphibian Monitoring Program.

Frontal dune(s): A ridge of unconsolidated sandy soil that extends along the shore but landward of the sand beach and seaward of the major dune formation. The profile of the frontal dune is defined by having relatively steep slopes that tend to have flatter and lower regions on each side of it.

Genera: More than one genus.

Genus: A group of species that are believed to be closely related because of their distinctive common characteristics.

GPS: A satellite-based navigation aid known as the Global Positioning System.

Gravid: A term in herpetology and other biological sciences to describe the condition of a female carrying eggs or young.

Gular: When applied to amphibians and reptiles this term is related to, or describing something located on, the outside of the throat.

Habitat: The natural home or environment of animals and plants.

Hectare(s): A metric unit equal to 2.471 acres.

Herbicide(s): A chemical substance that is toxic to plants and is used to control/destroy undesirable vegetation.

Herpetofauna: Collectively, the amphibians and reptile faunas of a given region.

Herpetology: The branch of zoology dealing with amphibians and reptiles.

Herps, or herptiles: Collectively, amphibians and reptiles. Herptile is a slang word which should be avoided.

Hydric habitat: A habitat characterized by moisture.

Hydrology: The science concerned with the earth's water, with emphasis on its movement in relation to the land.

Hydroperiod(s): The period of time during which a wetland holds water.

Hylid: Arboreal frogs of the family Hylidae, usually having toes that terminate in expanded disks; the tree frogs.

Impermeable: Not allowing fluid to pass through.

Immigrate: To come into a country or area.

Indigenous: Originating or occurring naturally in a particular place; a native species.

Insular: Living or located on an island.

Intergrade(s): When one subspecies merges with another where their ranges come together.

Intertidal zone: The area that is exposed to the air at low tide and underwater at high tide.

Invasive: An animal or plant that tends to spread prolifically and undesirably from a human perspective.

Jargon: Terminology which is used in relationship to a specific activity, profession, group, or event.

Kilometer(s): A metric unit of measurement equal to 1,000 meters (approximately 0.62 miles).

Labial(s): The labial scales are those that border the mouths of snakes and lizards. It can also refer to the papillae surrounding a tadpole's mouth.

Larvicide: A chemical used to kill insect pests in their larval stage.

Littoral: Waters that are part of, or along a shore, especially the seashore.

Macroalga: The larger aquatic photosynthetic plants that can be seen without the aid of a microscope.

Malacologist: The branch of invertebrate zoology which deals with the study of mollusks.

Malathion: A synthetic organophosphorous compound that is used as an insecticide.

Mean sea level (MSL): The level of the sea halfway between the mean levels of high and low water.

Mean high water (MHW): The highest average level that water reaches on a turning outgoing tide.

Mesic: Pertaining to a type of habitat having a medium level of moisture.

Meter(s): The basic unit of length in the metric system, equivalent to 39.37 U.S. inches.

Methoprene: A juvenile hormone analog that can be used as an insecticide and acts as a larval growth regulator in mosquito control.

Microburst(s): A sudden, powerful, localized downdraft of air.

Midden(s): A mound or deposit containing shells, animal bones, and other refuse that indicates the site of a human settlement; in this case the extinct Native Americans, the Calusa.

Migratory bird(s): A bird that singularly, or in large groups, makes regular seasonal long-distance journeys; e.g., cranes and waterfowl.

Millennium: A period of time equal to one thousand years.

Minute: Refers to small or tiny.

Mollusk(s): The preferred American spelling. The accepted British spelling is mollusc. The Mollusca are a large phylum of mainly marine, but also terrestrial, invertebrate animals, including snails and slugs.

Monel metal: An alloy comprised primarily of nickel and copper. It is highly resistant to corrosion.

Morphotype: Any of a group of different types of individuals of the same species in a population; a morph.

Mycoplasma: A pathogenic bacterial microorganism of the genus *Mycoplasma*.

Naled (Dibrom): A non-persistent organophosphate insecticide that is used to control mosquitoes.

Naris: A nostril; an external opening to the nasal cavity of a vertebrate.

Natal beach: The beach where a sea turtle hatched and to which a gravid female returns to deposit her eggs.

National Wildlife Refuge System: A national system of protected areas in the U.S. that are administered by the Fish and Wildlife Service of the Department of the Interior.

Natricine snake(s): Snakes placed in the family Natricidae by taxonomists. Eight common genera of U.S. snake species are classified as such, including the watersnakes and gartersnakes.

Neap tide: A tide that occurs when the difference between high and low tide is least; the lowest level of high tide. Neap tide comes twice a month, in the first and third quarters of the moon.

Neonate(s): Newborn; a term sometimes applied to amphibians and reptiles.

New Jersey light trap: An illuminated device that attracts and collects adult mosquitoes.

Nest site fidelity: The tendency for sea turtles and some freshwater species to return cyclically to the same geographic location for their successive nesting landings.

Niche: The status or functional role filled by an organism in a particular community.

Non-agrarian: A nonagricultural lifestyle.

Non-anuran amphibians: Newts and salamanders.

Nonindigenous: A plant or animal living outside of, or some distance beyond, its historic natural geographic range.

o/oo: The written symbol for parts per thousand.

Ocular: Pertaining to the eye; scales around the eye of snakes.

Organophosphate: An organic compound containing certain phosphates which may be toxic.

Ornithologist: A zoologist who specializes in the study of birds.

Pantropical: Distributed throughout the earth's tropical regions.

Parotoid gland(s): A pair of raised external glands atop the head of toads and spadefoots. These secrete variably poisonous bufotoxin for defense.

Parthenogenetic: Reproduction of an individual that results when there is no fertilization of a female by a male.

Pathogen: A bacterium, virus, or other microorganism that causes disease.

Pers. comm.: Abbreviation for personal communication.

Pioneer vegetation: In this use, plant species which colonize and thrive on the frontal dunes of beaches.

Pip(ped): To break, or to have broken through the eggshell.

PIT: Abbreviation for Passive Integrated Transponder, a tiny internally inserted tag; a microchip.

Plastron: The lower shell of a turtle.

Pleural scute(s): A series of large keratinized plates or scutes positioned laterally on each side of a hard-shelled turtle's carapace. They are located between the marginal scutes (those along the lower edge of the carapace) and the vertebral scutes that are aligned with and follow the length of the animal's backbone.

Pneumatophore(s): A root structure in certain plants, e.g., the mangroves, which serve as a respiratory organ.

Range: The geographical area in which a species normally occurs.

Range extension(s): The verification that a species inhabits an area outside of its currently known geographical range.

Recruitment: When juvenile organisms survive to be added as mature individuals to an established population.

Reptiles: An air-breathing, ectothermic (cold-blooded) vertebrate with an outer covering of scales or plates and possessing a bony skeleton, e.g., the alligator, a lizard, a snake, or a turtle.

Rookery: A concentrated breeding place of gregarious birds or sea turtles.

Rostral(s): The direction toward the end of the snout; the scale at the tip of a snake's snout.

Salinity: The concentration of salt in a solution.

Sanibel Slough: The interior freshwater wetland basin of Sanibel Island, Florida.

Scientific name(s): The Latin name given to an organism.

Sea strength: Referring here to salinity; the total amount of dissolved chlorides in grams in one kilogram of sea water; most of the Gulf of Mexico averages 28 to 32 parts per thousand (ppt).

Semidiurnal: Occurring during half a day; coming approximately once every 12 hours.

Semiterrestrial: To live mostly on land but requiring a moist environment, usually adjacent to nearby water.

Sympatrically: Where closely related species occupy the same or overlapping geographic areas without interbreeding.

Slider: Any of several freshwater turtles of the Genus, *Trachemys*.

Snout-to-vent-length (SVL): A measurement taken from the tip of the snout to the anal opening (the vent). Many amphibians and reptiles are measured using these standard points of reference, for accuracy and to compensate for often broken or disfigured tails in the case of lizards.

Spanish Land Grant: Typically, in colonial times, a gift of land from the Spanish crown to aristocrats or persons of cultural importance.

Species: A category of individuals that have common morphological characteristics and a high level of genetic similarity due to descending from a common ancestor.

Spoil pond(s): An excavated pond from which fill material is extricated; usually related to real estate development and highway construction.

Spring tide(s): A tide that occurs at the time of full or new moon; when high tides are higher.

Straightline or SCL: The shortest distance between two points; typically a measurement taken with calipers when a turtle carapace is being measured.

Stranded specimen: Typically, a marine creature that has washed ashore; these may be decomposed or still alive.

Study period: The length of time over which a study is conducted.

Subadult(s): An individual having adult characteristics but not yet sexually mature.

Subfamily: A taxonomic category that ranks below the family and above the genus.

Subsistence take: The harvest of wildlife for human subsistence; for survival.

Subspecies: A category in biological classification that ranks immediately below a species; a population of a particular geographic region morphologically distinguishable from other such populations of the same species, and is capable of interbreeding successfully with them where their ranges overlap.

Surface water aquifer: Ground water that occurs at or near the land surface; traditionally called the water table.

Taxonomy: The science of naming animals.

Terrapin: Any of a variety of aquatic turtles that are considered edible; the diamond-backed terrapins; the word is derived from the Native Americans, the Algonquian. In Europe, the term is applied to any aquatic, freshwater turtle. In the U.S., terrapin is associated with the brackish water turtles of the genus *Malaclemys*, and elsewhere several species of Asian *Batagur* are known as terrapins.

Thermal fog: A high-temperature technology used to produce large quantities of chemical fog without degrading the active pesticide ingredient.

Tidal amplitude(s): The vertical distance between high and low tides is the tidal range or amplitude.

Threatened: At risk of becoming endangered.

Tidal curve(s): A graphic representation depicting the rise and fall of the tide.

Toe pad(s): Arboreal frogs have enlarged pads, located on the ends of the toes, consisting of interlocking cells. When the climbing frog applies pressure to the toe pads, these cells maintain a grip through hydric suction which allows the frog to grip smooth surfaces like glass.

Tortoise: A herbivorous turtle that lives on land; e.g., the gopher tortoise. Tortoises have stumpy, powerful legs.

Tympanum: The often large external ear drum that is visible on the side of the head of most frogs behind their eyes.

Vegetational transition: Progressive natural (or post-developmental) transitional changes in vegetative communities over time; e.g., grasses to shrubs to conifers.

Ventral side (or venter): The underside of an organism.

Vocal sac(s): One of usually a pair of inflatable resonating sacs positioned in loose folds of skin on each side of the mouth of various frogs. Some species have only a single vocal pouch, such as toads and treefrogs.

Vouchered: Substantiating evidence; positive proof.

Voucher collection: Specimens, usually preserved, that serve as a basis of study and are retained in a collection as references. Vouchers also may consist of photographs, tissue samples, or DNA.

Watershed: High land that divides two areas that are being drained by different river systems.

West Indian hardwood hammock(s): An ecosystem consisting of broad-leafed trees, shrubs, and vines, nearly all of which are native to the West Indies.

Xeric habitat: Extremely dry habitat.

~NOTES~

Part I

Location (page 2)

[1] Caloosahatchee River, an accepted and long-published geographical name, is redundant and translates to "River of the Caloosa (*sic*) Calusa." Henceforth, herein we drop "River" and use Caloosahatchee singularly when mentioning this waterway.

Geologic Formation (page 3)

[2] This bridgeless barrier island is now commonly called Cayo Costa and is part of the state park system administered by the Florida Park Service.

[3] Hurricanes were not named by the U.S. National Weather Service until 1953.

Climate (page 5)

[4] Unlike the herpetofauna, in Florida manatees seek refuge against winter temperatures by migrating to the warmer water of natural springs and in the heated water discharged by power plants.

G. Bay Beach Zone (page 22)

[5] Where the system has not been so impacted by erosion and replaced by seawalls or other revetments or narrowed by encroachment of the Mangrove Zone.

Modification and degradation of aquatic habitat (page 25)

[6] The ditches dug by LCMCD were rarely more than 1.5 m (5 ft) below MSL and there was no de-watering during excavation.

Decline and recovery of the eastern indigo snake (page 27)

[7] The Orianne Society is a 501(c)(3) wildlife conservation organization that was founded in 2008. The goal of this non-profit organization is to save the eastern indigo snake.

A thoroughfare of major herpetofaunal mortality (page 28)

[8.] The COS Wildlife Committee was abolished on 31 October 2008.

Collection Techniques (page 31)

[9.] Among the most prominent of these during this period was Dr. R. Tucker Abbott, an American conchologist and malacologist.

Part II

An Overview (page 35)

[1.] Although a few of the SSAR-adopted common names used herein do not follow customary usage in Southwest Florida we adopt approved common names throughout this book to ensure our compliance with taxonomic trends and current nomenclature. We are consistent with the SSAR Committee on Standard English and Scientific Names. Where they exist, differing regionally-used common names for the herpetofauna of Southwest Florida are presented where appropriate.

CLASS AMPHIBIA (page 39)

[2.] The semi-aquatic dwarf salamander (*Eurycea quadridigitata*) ranges as far south as Sarasota County, in coastal Florida. There are recent reports of this species in inland Hendry County where, if the species is present, their habitat would be part of the Caloosahatchee watershed, but like the four listed genera this species could not have reached the barrier islands because of the salinity barrier.

Southern cricket frog (page 41)

[3.] We differ from the SSAR relative to this frog and have not adopted the subspecies, *Acris gryllus dorsalis*. According to Dodd (2013), there are two major clades within *A. gryllus*, an eastern clade occurring from the Florida Panhandle to southeastern Virginia and a western clade found in Mississippi and adjacent Tennessee (Gamble et al. 2008). The boundary between these clades is undetermined, but is likely in the Mobile Basin and Tombigbee River system. The distinction between these clades does not follow previous subspecific nomenclature (*A. g. gryllus* throughout most of the range and *A. g. dorsalis* in Florida). Despite some morphological differences (Neill 1950b), the recognition of *A. g. dorsalis* is not supported by molecular data.

Squirrel tree frog (page 56)

[4.] The pine woods treefrog (*Hyla femoralis*) occurs on the nearby mainland in flatwoods habitat, but has never been recorded on Sanibel or Captiva Island.

Cuban treefrog (page 58)

[5.] The barking treefrog (*Hyla gratiosa*) occurs on the nearby mainland; however, the species has never been recorded on Sanibel or Captiva Island.

Pig frog (page 66)

[6.] This national wildlife refuge was renamed in 1988, and is now known as Arthur R. Marshall Loxahatchee National Wildlife Refuge.

Southern leopard frog (pages 69 & 71)

[7.] These air temperature ranges are presented in F temperatures.

[8.] This frog is currently known by the common name of Northern Pacific treefrog.

[9.] Two species of freshwater crayfish, *Procambarus alleni* and *P. fallax*, occur on Sanibel Island. How these crayfish arrived on the island is unknown, but was likely because of early wildlife management efforts in the island's wetlands. We believe both species were indirectly but actively introduced. One possible way was that crayfish were transported with fingerling freshwater fish that were trucked from the Welaka National Fish Hatchery in Northeast Florida to Sanibel Island, in 1961 (LeBuff 1998). This shipment consisted predominantly of largemouth bass (*Micropterus salmoides floridanus*) and copperhead bream (*Lepomis macrochirus mystacalis*), both nonindigenous to Sanibel Island. Other native Florida freshwater fish species that have shown up on the island since 1961 may have been delivered mixed in with the hatchery stock as were the crayfish. Arguably, tadpoles have also been transported with hatchery-produced fish. The second suspected means of introduction is that juveniles of the other, or even both, crayfish species were introduced to the island a decade earlier. Crayfish may have been inadvertently collected along with plants from the Everglades that were transplanted in the BT. Circa 1950, common threesquare, *Scirpus pungens;* soft-stem bulrush, *Scirpus validus;* and banana water lily, *Nymphoides aquatica*, were introduced by the USFWS into the Sanibel Island wetlands from LNWR to establish the three plants as waterfowl food. Crayfish may have been "hitchhikers."

Loggerhead sea turtle (pages 79 & 82)

[10.] Twenty-one barnacle species are known from Northwest Atlantic loggerheads. The majority are obligate commensals of motile marine animals from the superfamily Coronuloidea. While most coronuloids will settle on nearly every surface of a host turtle, there are some species that are most common on only particular regions of turtles. Chelonibiine barnacles prefer to settle on the carapace while Platylepadid turtle barnacles are generally extra-carapacial, and prefer to settle on the skin. One of these, *Stomatolepas praegustator* preferentially settles within the buccopharyngeal region of loggerheads, and another, *Stephanolepas muricata*, preferentially settles on the leading edges of the front flippers. Two species, *Calyptolepas bjorndalae* and *Cylindrolepas darwiniana*, appear to be most common embedded in the marginal bones of the carapace.

Newer data on the barnacles of debilitated loggerheads indicates that turtle barnacles are generally harmless, but when turtles become immunosuppressed, barnacle colonization often overloads and deforms the host turtle—contributing to the turtle's decline, and possibly, the turtle's demise. (M. Frick, pers. comm.)

[11.] SI prefix tags were used primarily on study beaches in Brevard and Indian River counties, Florida, by Caretta Research, Inc. volunteers.

Green sea turtle (page 87)

[12.] Diseased turtles were specimens that were primarily affected by fibropapilloma tumors. The impacts of these tumors to individual green turtles varied. Some infections were severe enough that those turtles were immediately humanely euthanized because they were totally blinded by baseball-sized tumors and emaciated to the degree they were near death.

Leatherback sea turtle (pages 96 & 99)

[13.] Some herpetologists claim the estuarine crocodile (*Crocodylus porosus*) is heavier than the leatherback sea turtle, but most weights given for this crocodile are not reputable.

[14.] This length and weight are based on the record of a dead leatherback sea turtle from Wales, U.K., in 1988 (Morgan 1989).

[15.] The eggs of the giant tortoises of the Galapagos Islands equal those of the leatherback sea turtle in size, but those of the former are hard-shelled.

[16.] At hatching the leatherback sea turtles in E. J. Phillips' care averaged 6 cm (2.36 in) in CCL length. At release the largest measured 21.6 cm (8.5 in) CCL and weighed 1,035 g (2.28 lb).

Gopher tortoise (page 106)

[17.] The eggs of the gopher tortoise are usually only about one mm out of round.

Kemp's ridley sea turtle (pages 115 & 116)

[18.] At this writing, this carapace is in the custody of George Weymouth, Sopchoppy, Florida.

[19.] As an example, the long term effects of the 2010 Deepwater Horizon oil spill in the Gulf of Mexico may not be completely understood because of the lack of data on which to base recovery goals (Bjorndal et. al 2011).

Ornate diamond-backed terrapin (page 120)

[20.] We have been unsuccessful in locating the whereabouts of this photograph.

Red-eared slider (page 129)

[21.] The authors opted to consolidate the two members of the genus *Trachemys* into one account because of their similarities, and status as introduced species.

American alligator (pages 135, 136, & 140)

[22.] The publication was dated 1801, but was not actually published until 1802. The date of publication has the priority as the official date the name was given.

[23.] The spectacled caiman is native to South and Central America and was introduced to Florida as a result of the pet trade after American alligators and crocodiles became illegal to keep by the general public in the U.S.

[24.] This may not be true of all species, especially turtles. Some box turtles show no measureable growth over at least a 16 year period. Individuals may reach an asymptote in length, although they may continue to gain weight. (C. K. Dodd, Jr., pers. comm.)

[25.] Some borrow pits on the BT are an exception. They were dewatered during excavation until the practice was curtailed on Sanibel Island by water-use regulators.

[26.] Estero Island is also known as Fort Myers Beach.

American crocodile (pages 141, 142, & 144)

[27.] There have been recent published works where some popular Florida novelists and the press have inappropriately called this species "saltwater crocodile."

[28.] Other morphological dimensions for this Sanibel Island crocodile specimen are: SVL, 1.84 m (6.04 ft) and head length, 532 cm (17.45 in).

[29.] In the 1940s, one legendary crocodile hunter in the Florida Bay region, W. Argyle Hendry, regularly sold hatchling American crocodiles as "baby alligators" to tourists traveling on U.S. 1 near his home on Key Largo, in Monroe County.

West African rainbow lizard (page 150)

[30.] A lizard's length is correctly measured SVL, but in the following accounts we provide this dimension as total body length; from the tip of a perfect specimen's nose to the tip of its normal, undamaged tail.

[31.] University of Florida (UF) photographic voucher number: UF 146784.

Tokay gecko (page 163)

[32.] The largest gecko species is considered by herpetologists to be the New Caledonian giant gecko (*Rhacodactylus leachianus*). This gecko attains a SVL of 24.5 cm (9.6 in) and a total length of 35.6 cm (14 in), and regularly exceeds the tokay gecko in maximum length. New Caledonia is located in the southwest Pacific Ocean and is situated 1,500 km (930 mi) east of Australia.

Eastern indigo snake (pages 194 & 196)

[33.] The eastern indigo snake is now considered to be possibly extirpated in South Carolina (Hammerson 2007).

[34.] A walking trail/bicycle path now known as Indigo Trail follows what originally was a mosquito control dike on JNDDNWR. This trail is named for the indigo snake because the species was once regularly observed there.

Harlequin coralsnake (page 199)

[35.] The scarletsnake (*Cemophora coccinea*) is common on the mainland in some habitats, but as of this writing, this snake has not yet been recorded on Sanibel or Captiva islands.

Mangrove saltmarsh watersnake (page 205)

[36.] A specimen that serves as the original name bearer for a new species on which all others of that species are thereafter vouchered.

Florida cottonmouth (page 231)

[37.] The first was reported to LeBuff in 1969, and the second, supposedly observed in 1997, was reported to Lechowicz in 2002.

Boa constrictor (page 234)

[38.] The boa constrictor's scientific name is quite unique in herpetological nomenclature; its scientific and common names are exactly the same.

[39.] In 2012, boas and pythons that are known to be reproducing in Florida include: the boa constrictor, Burmese python (*Python molurus bivittatus*), and the African rock python (*Python sebae*). There is concern among wildlife biologists that other giant snakes, i.e., the green and yellow anacondas and the reticulated python may one day join their ranks.

Burmese python (page 235)

[40] According to the FFWCC this length represents the largest of this species ever recorded in Florida. A Burmese python of this length was killed in May 2013 in Southeast Dade County.

~SUGGESTED READING~

Anholt, B. 1998. *Sanibel's Story: Voices and Images from Calusa to Incorporation.* City of Sanibel. 191 pp.

Ashton, R. E., Jr., and P. S. Ashton. 1988. *Handbook of Reptiles and Amphibians of Florida: Part One: The Snakes.* Windward Publishing, Miami, Florida.

———. 1988. *Handbook of Reptiles and Amphibians of Florida: Part Two: The Lizards, Turtles & Crocodilians.* Windward Publishing, Miami, Florida.

———. 1988. *Handbook of Reptiles and Amphibians of Florida: Part Three: The Amphibians.* Windward Publishing, Miami, Florida.

———. 2004. *The Gopher Tortoise: A Life History.* Pineapple Press, Sarasota, Florida.

———. 2008. *The Natural History and Management of the Gopher Tortoise,* Gopherus polyphemus *(Daudin).* Krieger Publishing Company, Malabar, Florida.

Bartlett, R. D., and P. P. Bartlett. 1999. *A Field Guide to Florida Reptiles and Amphibians,* Gulf Publishing Company, Houston, Texas.

———. 2003. *Florida Snakes: A Guide to Their Identification and Habits*, University Press of Florida, Gainesville.

———. 2005. *Guide and Reference to the Snakes of Eastern and Central North America (North of Mexico)*, University Press of Florida, Gainesville.

———. 2006. *Guide and Reference to the Amphibians of Eastern and Central North America (North of Mexico)*, University Press of Florida, Gainesville.

———. 2006. *Guide and Reference to the Crocodilians, Turtles and Lizards of Eastern and Central North America (North of Mexico)*, University Press of Florida, Gainesville.

Behler, J. L., and F. W. King. 1979. *National Audubon Society: Field Guide to North American Reptiles and Amphibians*, Alfred A. Knopf, New York.

Buhlmann, K., T. Tuberville and W. Gibbons. 2008. *Turtles of the Southeast*, The University of Georgia Press, Athens.

Campbell, G. R. 1978. *The Nature of Things on Sanibel*. Press Printing Company, Fort Myers, Florida.

——— and A. L. Winterbotham. 1985. *Jaws Too!* Sutherlin Publishing, Fort Myers, Florida.

Conant, R. and J. T. Collins. 1998. *Peterson Field Guides: Reptiles & Amphibians, Eastern, Central, North America*, Third Edition, Revised. Houghton Mifflin Harcourt Publishing, Boston, Massachusetts.

Dodd, C. K., Jr. 2013. *Frogs of the United States and Canada*, 2 Volumes, Johns Hopkins University Press, Baltimore, Maryland.

Dorcas, M. and W. Gibbons. 2008. *Frogs and Toads of the Southeast*. The University of Georgia Press, Athens.

———and J. D. Willson. 2011. *Invasive Pythons in the United States*, University of Georgia Press, Athens.

———and E. M. Ernst. 2003. *Snakes of the United States and Canada*, Smithsonian Books, Washington, DC.

Ernst, C. H., and J. E. Lovich. 2009. *Turtles of the United States and Canada*, Second Edition. The John Hopkins University Press, Baltimore.

Gibbons, W., and M. Dorcas. 2005. *Snakes of the Southeast*, The University of Georgia Press, Athens.

———, J. Greene, and T. Mills. 2009. *Lizards and Crocodilians of the Southeast*, The University of Georgia Press, Athens.

Henderson, R. W., and R. Powell. 2007. *Biology of Boas and Pythons*, Eagle Mountain Publishing, Eagle Mountain, Utah.

Jensen, J. B., C. D. Camp, W. Gibbons, and M. J. Elliott. 2008. *Amphibians and Reptiles of Georgia*, The University of Georgia Press, Athens.

LeBuff, C. R., Jr. 1990. *The Loggerhead Turtle in the Eastern Gulf of Mexico*, Caretta Research, Inc. Sanibel, Florida.

———. 2011. Images of America: *J. N. "Ding" Darling National Wildlife Refuge*, Arcadia Publishing, Charleston, South Carolina.

Means, D. B. 2008. *Stalking the Plumed Serpent and Other Adventures in Herpetology*. Pineapple Press, Sarasota, Florida.

Meshaka, W. E., Jr. 2001. *The Cuban Treefrog in Florida*, University Press of Florida, Gainesville.

———, B. P. Butterfield and J. B. Hauge. 2004. *The Exotic Amphibians and Reptiles of Florida*, Krieger Publishing Company, Malabar, Florida.

———, and K. J. Babbitt. (eds) 2005. *Amphibians and Reptiles: Status and Conservation in Florida,* Krieger Publishing Company, Malabar, Florida.

Moler, P. E. 1992. *Rare and Endangered Biota of Florida: Volume III. Amphibians and Reptiles*, University Press of Florida, Gainesville.

Sobczak, C. 2010. *Living Sanibel: A Nature Guide to Sanibel and Captiva Islands,* Indigo Press L.L.C., Sanibel, Florida.

Tennant, A. 1997. *Gulf's Field Guide Series: A Field Guide to Snakes of Florida,* Gulf Publishing Company, Houston, Texas.

———. 2003. *Lone Star Field Guide: Snakes of Florida*, Taylor Trade Publishing, Lanham, Maryland.

Wheeler, D. G. 2001. *Tales from the Golden Age of Rattlesnake Hunting*, E.C.O. Publishing, Rodeo, New Mexico.

Whitney, E., D. B. Means, and A. Rudloe. 2004. *Priceless Florida. Natural Ecosystems and Native Species*, Pineapple Press, Sarasota, Florida.

Williams, W., and P. Carmichael. 2003. *Florida's Fabulous Reptiles and Amphibians,* World Publications, Tampa, Florida.

Wilson, L. D., and L. Porras. 1983. *The Ecological Impact of Man on the South Florida Herpetofauna*. University of Kansas Museum of Natural History, Special Publication No. 9, 89 pp.

~LITERATURE CITED~

Addison, D. S. 1996. *Caretta caretta* (Loggerhead Sea Turtle). Nesting Frequency. Herpetological Review 27:76.

Adler, K. 2004. America's first herpetological expedition: William Bartram's travels in Southeastern United States (1773-1777). Bonner Zoologische Beiträge 52:275-295.

Anholt, B. 1998. Sanibel's Story: Voices and Images from Calusa to Incorporation. City of Sanibel. 191 pp.

Barbour, T. 1944. That Vanishing Eden, A Naturalist's Florida. Little, Brown & Company, Boston.

Bartlett, R. D., & P. P. Bartlett. 1999. A Field Guide to Florida Reptiles and Amphibians. Gulf Publishing, Houston, Texas. 280 pp.

──────. 2006. Guide and Reference to the Crocodilians, Turtles, and Lizards of Eastern and Central North America (North of Mexico). University Press of Florida. Gainesville, Florida. 318 pp.

Beyer, S. M. 1993. Habitat Relations of Juvenile Gopher Tortoises and a Preliminary Report of Upper Respiratory Tract Disease (URTD) in Gopher Tortoises. M.S. Thesis, Iowa State University, Ames, Iowa. 95 pp.

Bjorndal, K. A., B. W. Bowen, M. Chaloupka, L. B. Crowder, S. S. Heppell, C. M. Jones, M. E. Lutcavage, D. Policansky, A. R. Solow, & B. E. Witherington. 2011. Better Science Needed for Restoration in the Gulf of Mexico. Science 331:537-538.

Blatchley, W. S. 1932. In Days Agone, Notes on the Fauna and Flora of Subtropical Florida in the Days When Most of Its Area was a Primeval Wilderness. The Nature Publishing Company, Indianapolis.

Boggess, D. H. 1974. The Shallow Fresh-Water System of Sanibel Island, Florida. Florida Department of Natural Resources, Bureau of Geology. Report of Investigations 69. 52 pp.

Boykin, C. S. 2005. The Status and Demography of the Ornate Diamondback Terrapin (*Malaclemys terrapin macrospilota*) within the Saint Martins Marsh Aquatic Preserve. Unpublished report to Florida Department of Environmental Protection. Miami, Florida.

Brown, M. B., G. S. McLaughlin, P. A. Klein, B. C. Crenshaw, I. M. Schumacher, D. R. Brown, & E. R. Jacobson. 1999. Upper Respiratory Tract Disease in the Gopher Tortoise Is Caused by *Mycoplasma agassizii*. American Society for Microbiology. 37:2262–2269.

Butler, J. A., & G. L. Heinrich. 2007. The Effectiveness of Bycatch Reduction Devices on Crab Pots at Reducing Capture and Mortality of Diamondback Terrapins *(Malaclemys terrapin)* in Florida. Estuaries and Coasts 30:179–185.

Byrne, M. J., & J. N. Gabaldon. 2008. Hydrodynamic Characteristics and Salinity Patterns in Estero Bay, Lee County, Florida: U.S. Geological Survey, Scientific Investigations Report 2007-5217, 33 pp.

Campbell, G. R. 1978. The Nature of Things on Sanibel. Pineapple Press, Sarasota, Florida. 176 pp.

──────. 1985. "Much Ado About Nothing". Sanibel-Captiva Islander. August 12, 1985. p. 1B.

──────. 1985a. The Nocturnal "Glee Club" Revisited. Island Reporter. September 24, 1985. p. 1B.

Campbell, G. R. & A. L. Winterbotham. 1985. Jaws, Too! The Natural History of Crocodilians With Emphasis on Sanibel Island's Alligators. Sutherland Publishing, Fort Myers, Florida. 268 pp.

Carr, A. F., Jr. 1940. A Contribution to the Herpetology of Florida. University of Florida Publications, Biological Science Series. 3:1–118.

Clark, J. R. 1976. The Sanibel Report. The Conservation Foundation, Washington DC. 305 pp.

Clark, J. R., P. W. Borthwick, L. R. Goodman, J. M. Patrick, Jr., E. M. Lores, & J. C. Moore. 1987. Effects of Aerial Thermal Fog Applications of Fenthion on Caged Pink Shrimp, Mysids and Sheepshead Minnows. American Mosquito Control Association 3:466-72.

Cooley, G. R. 1955. The Vegetation of Sanibel Island Lee County, Florida. Rhodora 57:1-32.

Crother, B. I. (compiler). 2008. Scientific and Standard English Names of Amphibians and Reptiles of North America North of Mexico. SSAR Herpetological Circular 37.

Cushing, F. H. 1896. Exploration of Ancient Key Dwellers' Remains of the Gulf Coast of Florida. American Philosophical Society Proceedings 35 (1897): 329-448. Reprinted, 2001. University Press of Florida, Gainesville.

Davenport, J. 1997. Temperature and the Life-History Strategies of Sea Turtles. Journal of Thermal Biology 22:479-488.

Dodd, C. K., Jr. 2013. Frogs of the United States and Canada, 2 Volumes, Johns Hopkins University Press, Baltimore, Maryland.

Duellman, W. E., & A. Schwartz. 1958. Amphibians and Reptiles of Southern Florida. Bulletin of the Florida State Museum, Biological Sciences. 3:181–324.

Dundee, H. A., & D. A. Rossman. 1989. Amphibians and Reptiles of Louisiana. Louisiana State University Press. Baton Rouge, Louisiana. 300 pp.

Enge, K. M., K. L. Krysko & B. L. Talley. 2004. Distribution and Ecology of the Introduced African Rainbow Lizard, *Agama agama africana* (Sauria: AGAMIDAE) in Florida. Florida Scientist 67:303-310.

Ehrhart, L. M. 2011. Where Sea Turtles Nest. St. Petersburg Times. September 18, 2011.

Ernst, C. H., J. E. Lovich, & R. W. Barbour. 1994. Turtles of the United States and Canada. Smithsonian Institution Press, Washington, DC. 578 pp.

Gamble, T., P. B. Berendzen, H. B. Shaffer, D. E. Starkey, & A. M. Simons. 2008. Species Limits and Phylogeography of North American Cricket Frogs (*Acris*: Hylidae). Molecular Phylogenetics and Evolution 48:112-125.

Green, D. E., & C. K. Dodd, Jr. 2007. Presence of Amphibian Chytrid Fungus Batrachochytrium dendrobatidis and Other Amphibian Pathogens at Warmwater Fish Hatcheries in Southeastern North America. Herpetological Conservation and Biology 2:43-47.

Hammerson, G. A. 2007. *Drymarchon couperi*. In: IUCN 2012. IUCN Red List of Threatened Species. Version 2012.1.

Hansknecht, K. A. 2008. Lingual Luring by Mangrove Saltmarsh Snakes (*Nerodia clarkii compressicauda*). Journal of Herpetology 42:9–15.

Henderson, D. M. 2003. Effects of Stomach Stones on the Buoyancy and Equilibrium of a Floating Crocodilian: a Computational Analysis. *Canadian Journal of Zoology* 81:1346-1357.

Jensen, J., C. D. Camp, J. W. Gibbons, & M. Elliot (eds.). 2008. Amphibians and Reptiles of Georgia. University of Georgia Press, Athens.

Johnson, W. R. 1952. Range of *Malaclemys terrapin rhizophorarum on the West Coast of Florida*. Herpetologica 8:100.

Karlin, M. L. 2008. Distribution of *Mycoplasma agassizii* in a Gopher Tortoise Population in South Florida. Southeastern Naturalist 7:145-158.

Klinkenberg, J. 2008. Crocodile Making a Comeback in Florida. St. Petersburg Times. April 9, 2012.

Lamb, T. 1984. The Influence of Sex and Breeding Conditions on Microhabitat Selection and Diet in the Pig Frog, *Rana grylio*. American Midland Naturalist 111:311–318.

LeBuff, C. R., Jr. 1957. The Range of *Crocodylus acutus* Along the Florida Gulf Coast. Herpetologica 13:188.

—————. 1959. Quarterly Narrative Report (IV), Sanibel National Wildlife Refuge.

—————. 1969. The Marine Turtles of Sanibel and Captiva Islands, Florida. Sanibel-Captiva Conservation Foundation, Special Publication No. 1. 14 pp.

LeBuff, C. R., Jr., & R. W. Beatty. 1971. Some Aspects of Nesting of the Loggerhead Turtle, *Caretta caretta caretta* (Linne), on the Gulf Coast of Florida. Herpetologica 27:153–156.

LeBuff, C. R., Jr. 1974. Unusual Nesting Relocation in the Loggerhead Turtle, *Caretta caretta*. Herpetologica 30:29–31.

―――――. 1990. The Loggerhead Turtle in the Eastern Gulf of Mexico. Caretta Research, Inc., Sanibel Island, Florida. 216 pp.

―――――. 1998. Sanybel Light. Amber Publishing, Fort Myers, Florida. 304 pp.

―――――. 2007. Crocodile Comeback—The Sanibel Connection. Southwest Florida Natural History Newsletter 1:1.

―――――. 2010. Everglades Wildlife Barons: The Legendary Piper Brothers and Their Wonder Gardens. Ralph Curtis Publishing, Sanibel, Florida. 272 pp.

―――――. 2011. Images of America: J. N. "Ding" Darling National Wildlife Refuge. Arcadia Publishing, Charleston, South Carolina.

Lechowicz, C. 2009. Amphibians and Reptiles of Sanibel Island, Sanibel-Captiva Conservation Foundation. 1 p.

Lever, C. 2003. Naturalized Reptiles and Amphibians of the World. Oxford University Press, Oxford, 318 pp.

Lillywhite, H. B. and R. J. R. McCleary. 2008. Trophic Ecology of Insular Cottonmouth Snakes: Review and Perspective. South American Journal of Herpetology 3:175-185.

Martof, B. 1954. Variation in a Large Litter of Gartersnakes, *Thamnophis sirtalis sirtalis*. Copeia. 1954:2, 100-105.

McAllister, K. R., W. P Leonard, D. W. Hays, & R. C. Friesz. 1999. Washington State Status Report for the Northern Leopard Frog. Washington Department of Fish and Wildlife, Olympia, 36 pp.

McCoy, C. J., Jr. 1972. *Hemidactylus garnotii*. Herpetological Review 4:23.

McIlhenny, E. A. 1935. The Alligator's Life History. Boston, Massachusetts.

McLaughlin, G. S., E. R. Jacobson, D. R. Brown, C. E. McKenna, I. M. Schumacher, H. P. Adams, M. B. Brown, & P. A. Klein. 2000. Pathology of Upper Respiratory Tract Disease of Gopher Tortoises in Florida. Journal of Wildlife Diseases, 36:272–283.

McPherson, B. F., R. T Montgomery, & E. E. Emmons. 2007. Phytoplankton Productivity and Biomass in the Charlotte Harbor Estuarine System, Florida. Journal of the American Water Resources Association. 26:787-800.

Meshaka, W. E. Jr. 2001. The Cuban Treefrog in Florida. University Press of Florida, Gainesville.

Missimer, Thomas. 1973. Growth Rates of Beach Ridges on Sanibel Island, Florida. Transactions of the Gulf Coast Association of Geological Societies. 23:388-393.

Moriarty, E. C. & D.C. Cannatella. 2004. Phylogenetic Relationships of the North American Chorus Frogs (*Pseudacris*: Hylidae). Phylogenetics and Evolution 30:409–420.

Morgan, P. J. 1989. Occurrence of Leatherback Turtles *Dermochelys coriacea* in the British Isles in 1988 With Reference to a Record Specimen. Pages 119-120 in S. A. Eckert, K. L. Eckert, & T. H. Richardson, compilers. Proceedings of the Ninth Annual Conference on Sea Turtle Conservation and Biology. NOAA Technical Memorandum NMFS-SEFC-232. U.S. Fish and Wildlife Service, Vero Beach, Florida.

Morton, J. F. 1980. The Australian pine or beefwood (*Casuarina equisetifolia* L.), an invasive "weed" in Florida. Proceedings Florida State Horticultural Society.

Neil, W. T. 1971. The Last of the Ruling Reptiles, Alligators, Crocodiles and Their Kin. Columbia University Press, New York.

Ortenburger, A. I. 1928. The Whip Snakes and Racers: Genera *Masticophis* and *Coluber*. Memoirs University of Michigan. Museum, Ann Arbor, Michigan 1:1-247.

Phillips, E. J. 1977. Raising Hatchlings of the Leatherback Turtle. British Journal of Herpetology 5:677–678.

Pritchard, P. C. H. 1971. The leatherback or leathery luth, *Dermochelys coriacea*. IUCN Monograph 1, 39 pp.

Provost, M. W. 1953. The Water Table on Sanibel Island. Florida State Board of Health, Bureau of Sanitary Engineering. Division of Entomology, Vero Beach, Florida.

Rice, K. G., J. H. Waddle, M. W. Miller, M. E. Crockett, F. J. Mazzotti, and H. F. Percival. 2011. Recovery of native treefrogs after removal of nonindigenous Cuban treefrogs, *Osteopilus septentrionalis*. Herpetologica 67:105-117.

Rizkalla, C. E. 2009. First Reported Detection of *Batrachochytrium dendrobatidis* in Florida, USA. Herpetological Review 40:189-190.

Rothermel, B. B., S. C. Walls, J. C. Mitchell, C. K. Dodd, Jr., L. K. Irwin, D. E. Green, V. M. Vazquez, J. W. Petranka, & D. J. Stevenson. 2008. Widespread Occurrence of the Amphibian Chytrid Fungus *Batrachochytrium dendrobatidis* in Amphibian Populations in the Southeastern USA. Diseases of Aquatic Organisms 82:3-18.

Sheng, Y. P. 1996. Circulation in the Charlotte Harbor Estuarine System. South Florida Water Management District, West Palm Beach, Florida.

Simpson, C. T. 1932. Florida Wild Life. The MacMillan Company, New York.

Small, J. K. 1929. From Eden to Sahara, Florida's Tragedy. The Science Press Printing Company, Lancaster, Pennsylvania.

Sobczak, C. 2006. Alligators, Sharks and Panthers: Deadly Encounters with Florida's Top Predator—Man. Indigo Press, Sanibel, Florida. 336 pp.

Solecki, W. D. et al. 1999. Human–Environment Interactions in South Florida's Everglades Region: Systems of Ecological Degradation and Restoration. Urban Ecosystems. Springer, Netherlands. 3:3-4.

Sparling, D. W., T. P. Lowe, & A .E. Pinkney. 1997. Toxicity of Abate® to Green Frog Tadpoles. Bulletin of Environmental Contamination and Toxicology 58:475–481.

Sparling, D. W. & G. M. Fellers. 2009. Toxicity of Two Insecticides to California, USA, Anurans and its Relevance to Declining Amphibian Populations. Environmental Toxicology and Chemistry 28:1696–1703.

Stebbins, R. C. & N. W. Cohen. 1995. A Natural History of Amphibians. Princeton University Press, Princeton, New Jersey. 316 pp.

Truitt, J. O. 1962. A Guide to the Snakes of South Florida. Privately Published. 46 pp.

Tucker, A. D. 2009. Eight Nests Recorded for a Loggerhead Turtle Within One Season. Marine Turtle Newsletter 124:16-17.

Wang, J. C. S., & Raney, E. C. 1971. Distribution and Fluctuations in the Fish Fauna of the Charlotte Harbor Estuary, Florida. Mote Marine Laboratory. 56 pp.

Wilson, L. D., & L. Porras. 1983. The Ecological Impact of Man on the South Florida Herpetofauna. University of Kansas Museum of Natural History. Special Publication 9:1-89.

~ABOUT THE AUTHORS AND PHOTOGRAPHERS~

Charles LeBuff received specialized training in herpetology as an intern at the Boston Museum of Science and through related coursework at Tufts College (now University) in his hometown of Medford, Massachusetts. Before his family moved to Bonita Springs, Florida, in 1952, he received advanced technical training in curatorial and systematic herpetology at the Museum of Comparative Zoology at Harvard University, where he was mentored by the museum's curator of herpetology, Arthur Loveridge. LeBuff published his first article in the scientific journal *Herpetologica* at age 14 and has since written a number of scientific papers and books, many related to amphibians and reptiles.

LeBuff studied the herpetofauna of the region around Naples, Florida, and soon developed an interest in the biology and conservation of sea turtles. He is widely recognized as a conservationist and a pioneer in Florida sea turtle biology.

In 1956, LeBuff accepted a position with the USFWS Red Tide Investigation Field Station in Naples. During this time he continued his long-term studies of South Florida herpetology, and in 1957 he was included in a list of herpetological authorities in the book, *Handbook of Snakes*.

In 1958, LeBuff transferred to the SNWR, where he continued his independent studies of crocodiles and sea turtles.

From 1963 until 1966, LeBuff represented the Fish and Wildlife Service as a member of the American Alligator Council, a group of alligator experts charged with developing methodologies of how best to protect dwindling American alligator populations.

LeBuff founded the non-profit sea turtle conservation organization Caretta Research, Inc., in 1968, and in 1971 redirected his herpetological interests to this specialized field. In 1972 he was issued Marine Turtle Permit-001 by the State of Florida, which he held until 2012. His sea turtle program on the islands was transferred to the SCCF in 1991. As of 2013, the Sanibel and Captiva Islands sea turtle program has operated continuously every summer for 54 consecutive years, one of the longest continuously operating sea turtle monitoring programs in the world.

LeBuff was elected to the City of Sanibel's first city council in 1974. During his six-year political career, he championed environmentally focused ordinances and oversaw their implementation. Among these ordinances were land use regulations that helped protect Sanibel Island's vital ecosystems, regulated interactions between people and American alligators, and reduced negative human-related impacts on nesting sea turtles by the promulgation of some of the first beach lighting regulations.

LeBuff and his wife, Jean, now live in the Fort Myers area where LeBuff enjoys reading, writing, woodcarving, and dabbling in genealogy. He still frequently gets out into the field to look for herpetofauna. On rainy summer nights it is not unusual to find him slowly cruising rural roadways listening for and identifying the species of noisy frogs and toads from their loud and distinctive vocalizations. At the same time, he continues to search roads that pass through flooded wetlands for two species of South Florida snakes that he has never observed alive in the wild, even after more than 60 years of looking for them. These are the South Florida swampsnake (*Seminatrix pygaea paludis*) and the peninsula crowned snake (*Tantilla relicta relicta*). Annual summertime visits to Florida's best sea turtle nesting beaches also remain an important part of his life.

Chris Lechowicz began his study of amphibians and reptiles in Chicago. His interest began in second grade with a 10-gallon fish tank and a copy of the *Golden Guide to Reptiles and Amphibians*. His love of the aquatic environment soon transitioned into frogs, newts, and turtles. Growing up near the railroad tracks led to countless expeditions along the tracks to find Chicago gartersnakes. This blossomed into pseudo-fishing at the local lagoon where he concentrated his effort on turtles. At his father's urging, he tried golf in eighth grade, but much to his dad's chagrin, Lechowicz would purposely hit the ball into the water hazard so he could look for frogs, snakes, and turtles among the aquatic vegetation.

In his teenage years, Lechowicz discovered the CHS where he was mentored by many local and regional herpetologists. He helped numerous researchers, graduate students, and professors with their fieldwork during his teenage years into his 20s. Through the CHS he went on numerous herp excursions around the country. It was a trip to the Southeast with his future mentor, Ron Humbert, where Lechowicz found his favorite group of animals, the map turtles (the genus *Graptemys*). He has been actively studying this genus ever since and developed a natural history and resource website, Graptemys.com, in 2001.

Lechowicz attended Southern Illinois University at Carbondale where he received bachelor of science degrees in zoology and computer science. At this time, he is completing his thesis for a master's of science in environmental science at Florida Gulf Coast University. His research involves the map turtle species found in the Choctawhatchee River in northern Florida and southern Alabama.

In 2002, Lechowicz was hired by the SCCF to be its biologist/herpetologist. He conducts annual monitoring of Sanibel's herpetofauna and keeps the island amphibian and reptile list current. In 2005-2006, Lechowicz was funded to conduct tortoise research in Madagascar on the four endemic tortoises. He was promoted to director of the wildlife habitat management program at SCCF in 2011.

Lechowicz is an active member of the Calusa Herpetological Society, the Chicago Herpetological Society, Turtle Survival Alliance, Florida Turtle Conservation Trust, Diamondback Terrapin Working Group, and the Southwest Florida CISMA (Cooperative Invasive Species Management Area). He is also in a Fort Myers rock band called Vernal Pool.

Bill Love has been fascinated by amphibians and reptiles since his childhood in New Jersey, then for 40 years in Florida. His passion is photographing these animals in the wild.

He frequently gives educational presentations, leads eco-tours to Madagascar via Blue Chameleon Ventures (also Florida and Arizona, via ReptileRally.com), writes freelance articles on various herpetological subjects, and writes the monthly "Herpetological Queries" column in *REPTILES Magazine*.

Love has bred numerous species in captivity (creating many prominent morphs popular among collectors today), propagated snakes on a commercial scale, imported/exported them worldwide as co-owner of Glades Herp, Inc., and traveled worldwide observing and recording their natural histories digitally and on film. His imagery has appeared in books, magazines, calendars, business promotions, television, and websites for more than four decades.

Daniel Parker is the owner of Sunshine Serpents and a biologist for the University of Central Florida. His company is involved with photography, film, wildlife tours, education, and captive breeding of reptiles and amphibians. Parker's photographs have appeared in *Reptiles*, *Reptilia*, and *Herp Nation* magazines, as well as numerous publications worldwide.

His biological interests include eastern indigo snakes, gopher tortoises, sand skinks, Florida black bears, and many established exotics such as Burmese pythons. Parker hopes his work has helped promote an increased understanding of Florida's natural environment and the wildlife that lives there.

~INDEX, SCIENTIFIC NAMES~
(Herpetofauna only)

Acris crepitans 42
Acris gryllus 41, 61
Agama agama africana 149
Agkistrodon piscivorus conanti 229
Alligator mississippiensis 20, 135
Alligator sinensis 135
Amphiuma means 39
Anaxyrus quercicus 44
Anaxyrus terrestris 14, 45, 47
Anolis carolinensis 20, 151
Anolis equestris equestris 153
Anolis sagrei 155, 185
Apalone ferox 75, 126
Aspidoscelis sexlineata 14, 157

Basiliscus vittatus 159
Boa constrictor 233

Caiman crocodilis 135, 141
Caretta caretta 14, 78
Cemophora coccinea 199, 255
Chelonia mydas 14, 84
Chelydra serpentina 89
Coluber constrictor priapus 183
Coluber flagellun flagellum 14, 185

Crocodylus acutus 5, 20, 141
Crocodylus porosus 253
Crotalus adamanteus 14, 188

Deirochelys reticularia chrysea 92
Dermochelys coriacea 14, 95
Diadophis punctatus punctatus 192
Drymarchon corais couperi 195
Drymarchon couperi 14, 194

Eleutherodactylus planirostris 49
Eretmochelys imbricata 100
Eurycea quadridigitata 252

Gastrophryne carolinensis 51
Gekko gecko 162
Gopherus polyphemus 14, 18, 105

Hemidactylus garnotii 164
Hemidactylus mabouia 166
Hyla cinerea 53, 56
Hyla femoralis 252
Hyla gratiosa 58, 252
Hyla squirella 20, 55

Iguana iguana 168

Kinosternon baurii 109
Kinosternon subrubrum steindachneri 32, 111

Lampropeltis elapsoides 197, 199
Leiocephalus carinatus armouri 171
Lepidochelys kempii 14, 114
Lepidochelys olivacea 115
Lithobates catesbeianus 65
Lithobates clamitans 70
Lithobates grylio 37, 64, 71
Lithobates sphenocephalus 68

Macrochelys temminckii 91
Malaclemys terrapin macrospilota 14, 20, 118
Malaclemys t. rhizophorarum 120
Micrurus fulvius 179, 199
Micrurus tener 199

Nerodia clarkii compressicauda 20, 204
Nerodia fasciata pictiventris 204, 206, 231
Nerodia floridana 209
Nerodia taxispilota 211
Notophthalmus viridescens piaropicola 39

Ophisaurus ventralis 173
Osteopilus septentrionalis 20, 57, 185

Pantherophis alleghaniensis 20, 214
Pantherophis guttatus 217
Plestiodon inexpectatus 175
Pseudacris nigrita 60
Pseudacris ocularis 62
Pseudemys c. suwanniensis 126
Pseudemys nelsoni 122
Pseudemys peninsularis 124

Pseudobranchus axanthus belli 39
Python molurus bivittatus 235, 255
Python sebae 255

Rhinella marina 72
Ramphotyphlops braminus 220

Scincella lateralis 178
Siren lacertina 39
Sistrurus miliarius barbouri 232
Storeria victa 222

Terrapene carolina bauri 14, 127
Thamnophis sauritus sackenii 224
Thamnophis sirtalis sirtalis 226
Trachemys scripta elegans 131
Trachemys scripta scripta 130

Varanus niloticus 180

~INDEX, GENERAL~

500-year storm event 239

Abate 70, 239
Abbott, R. Tucker 252
active introduction 36
adulticide 239
advertisement sounds 239
Aedes taeniorhynchus 11
Allen, William 6
alligator,
    harvest policy 138
    illegal harvest 139
    policy, City of Sanibel 140
    snapping turtle 89, 91
    surveys 32, 138
    tagging 139
    vocalization 137
    American xvi, 10, 17, 20, 30, 123, 135
    Chinese 135
Alligator Trapper 239
American,
    bullfrog 65
    chameleon 152
    crocodile 5, 20, 141
    Society of Ichthyologists and Herpetologists 35
    toad 48

anal warts 239
anerythristic 219
Anholt, Betty xviii, 30
anole,
    brown 152, 155, 185
    Cuban 153
    green 151
    western knight 153
aquatic turtle sampling 31
Archie Carr National Wildlife Refuge 80
arribada 115, 240
artesian wells 11
asexual 240
Australian pine 202
Australian pines 179
Averill, Bob 134
Avicennia germinans 19, 20

Bailey Tract 4, 10, 17, 25, 254
Bailey, Frank P. 10
Baldwin, Theresa 146
barnacles 79, 253
Bartram, William xi
basilisk, brown 159
bastard turtle 116
Batrachochytrium dendrobatidis 71

271

Bay Beach Zone 22
Beach Road Weir 43
beachfront development 86
Beatty, Richard xvii, 82
Big Cypress Basin 231
biogenic 240
biological diversity 240
black mangroves 19
Blind Pass 3, 13
blindsnake, brahminy 221
boa constrictor 234
bobcat, Florida 191
Boerema, Michael 120
Boutchia, Warren xvii
brahminy blindsnake 220
Brazilian pepper 17, 108, 203
brownsnake, Florida 223
Bryant, Amanda xviii, xix, 94
Bti 70
bufotoxin 240
bullfrog 65
Burmese python 234, 235
bycatch reduction device 120

caiman, spectacled 135, 141
Caloosahatchee 2, 8, 23, 24, 39, 251
Calusa 6
Campbell, George 31, 74, 129, 136, 138, 139, 236
Campbell, Todd 181
cane toad 73
canine distemper 81
Caouette, Joel xix, 124
Cappiolla, Stephanie 74
Captiva Erosion Prevention District 4
Caretta Research, Inc. xvii, 82, 98, 253
Carr, Archie xi, 82
caruncle 106

Casuarina equisetifolia 179, 202
caudal scute 240
Cayo Costa 27, 251
Charlotte Harbor Estuarine System 2, 103, 142
Chinese alligator 135
chytrid fungus 241
City of Sanibel 5, 14, 24, 28, 81, 140, 159, 170
Clam Bayou 13
Clark, John 12
Clark, Toby xix
Clinic for the Rehabilitation of Wildlife 28, 103, 175
Coastal Construction Control Line 12
coastal plain 240
Collier County Herpetological Society 213
Comprehensive Land Use Plan 7
Conocarpus erectus 17
controlled burns 202
Cook, Cassie 74
Cooley, George R. 2, 12
cooter,
    Florida red-bellied 123
    Peninsula 126
coralsnake,
    harlequin 201
    Texas 199
cottonmouth 212, 229, 230
cottontail rabbit 189
coyote 81
cranial crest 73, 241
crayfish 71, 253
critter crossings 28
crocodile,
    American 141
    estuarine 253
Cuban treefrog 55, 57
Curtis, Ralph xvii

Darling, J. N. "Ding" xxi, 11, 196
DDT 11
Decline and recovery of the eastern indigo snake 26
Deepwater Horizon 254
detiologic agent 241
diapause 110, 112
dime-store turtle 133
dog-bone groins 4
dorsolateral ridge 242
drift fence 242
drift-fence collecting 31
dusky pigmy rattlesnake 232

eastern,
    coachwhip 185
    diamond-backed rattlesnake 189
    gartersnake 227
    glass lizard 173
    gray squirrel 189
    indigo snake 26, 30, 187, 195
    narrow-mouthed toad 51
    ratsnake 214
eelgrass 85
elapid 242
embryonic diapause 93
Endangered Species Act 82, 98
Endangered Species List 139, 144
endemic 242
epibionts 101, 242
Estero Bay 6
Estero Island 140
Everglades National Park 29, 59, 142, 144
Everglades Wonder Gardens 200
exotic pest plants xvi

Faillace, Cara xix, 236
Fannie's Preserve 185

Farallo, Vinny xix
Federal Register 8
Fenthion 70
fibropapilloma tumors 86, 253
fire ant 81, 129
flooding 7, 16
Florida,
    bobcat 191
    box turtle 32, 127
    brown watersnak 212
    brownsnake 223
    chicken turtle 92
    cottonmouth 207
    cricket frog 71
    green watersnake 209
    kingsnake 30
    Land Company 6
    manatee 5
    marsh rabbit 189
    mud turtle 25, 32, 111
    record alligator 136
    red-bellied cooter 122
    scarletsnake 199, 255
    softshell turtle 76
    watersnake 204, 207, 231
    Fish and Wildlife Conservation Commission 98, 136, 147, 236
    Game and Freshwater Fish Commission 7, 139, 141
    Museum of Natural History 117
    Panther National Wildlife Refuge 29
    State Board of Health 11, 44
    Florida Trustees of the Internal Improvement Trust Fund 7
    Wildlife Code 141
Fort Myers News-Press 30
fossorial species 243
Fox, Stanley xviii

Frannie's Preserve 219
Frog Watch Network 31, 243
frog-call surveys 31
frog,
- green 70
- greenhouse 50
- little grass 63
- northern leopard 70
- Pacific chorus 71
- pig 65
- southern chorus 61
- southern cricket 41
- southern leopard 69

Frommer, Arthur 2
Fuller, Sam xix

gartersnake, eastern 226
gastroliths 137
Gator Heaven 66, 138
gecko,
- giant 255
- Indo-Pacific house 164
- tokay 162
- tropical 167
- wood slave 166

genus 36
ghost crab 81
Global Positioning System 32
Gopher Derby 108
gopher tortoise 18, 32, 105
gopher tortoise burrows 32, 107
green,
- iguana xvi
- sea turtle 84
- treefrog 53

greenhouse frog 49
Gulf Beach Ridge Zone 10, 14, 159, 186

Gulf Beach Zone 12
gulf water temperatures 5
gulfweed 80
habitat,
- enhancement 94
- fragmentation 208
- restoration 27

harlequin coralsnake 199
Harpham, Malcolm 151
hawksbill sea turtle 100
hectare 244
hemotoxic venom 230
Henahan, Tony 74
Hendry, W. Argyle 255
Herpetologists' League 35
herpetology 1
Herrera, Andres 116
hispid cotton rat 189
Holly, Susie xviii
House Bill 1095 7
Humbert, Ron xviii
Hurricane,
- Charley 5, 26, 151, 179
- Donna 5, 19, 62, 202

hypothermic stunning 103

iguana, green 168
impacts from landscaping 26
impacts of fire 26
incidental catch 102
indigo snake, reintroduction 27
Indigo Trail 255
Inter-Island Conservation Association 7
intergrade 134
Interior Wetland Basin Zone 15

J. N. "Ding" Darling National Wildlife Refuge xvii, 7, 17, 70, 117, 136, 140, 142, 145, 170, 192, 213

Jesus Christ lizard 161

Johnson, Shane xix

Jones, Lennie 31, 213

Keewaydin Island 82

Kemp's ridley sea turtle 78, 114

kingsnake,
    Florida 30
    scarlet 198

Kirkland, Natalie xvii

LaCosta Island 3

Laguncularia racemosa 19, 20

Lake Okeechobee 2, 8, 23

land acquisition 28

Landry, Sam 173

Las Conchas Road 43

leatherback sea turtle 95

LeBuff,
    Jean xvii
    Laban xvii
    Laurence xvii

Lechowicz, Nicole xviii

Lee County 4, 7, 25

Lee County,
    Hyacinth Control District 24
    Mosquito Control District 11, 25, 70, 94, 251
    Refuge Commission 7

lightning strikes 10

Lindblad, Erick xviii

Linnaeus, Carl 35

little grass frog 62

live oak 202

lizard, eastern glass 175

lizard,
    northern curly-tailed 172
    West African rainbow 150

Loflin, Rob 192

loggerhead sea turtle 30, 78

loggerhead turtle, CR-140 82

Love, Bill xix, 267

Loxahatchee National Wildlife Refuge 66, 252

Lynx rufus floridanus 191

Malathion 11, 71

manatee grass 85

Mangrove Head 10, 17, 25

Mangrove Zone 18, 205

Marco Island 142

Marine, "Sonny" xvii

maximum air temperatures 5

Mazzotti, Frank 145

McCoy, Clarence J. xxi, 166

McIntyre Creek, 145

Mean High Water 13

Mean Sea Level 14

melanistic 227

metric system 36

Mid-Island Ridge Zone 17

Miller, T. Wayne 11

modification and degradation of aquatic habitat 24

Monel metal tags 82, 245

monitor, Nile 181

morphotype 245

Mound Key 6

multiple nesting 97

Mycoplasma agassizii 108

Naled 70

National,
    Key Deer Refuge 29
    Wilderness System 202

Wildlife Refuge System 7
natricine snakes 212
Nave, Allen 191
neurotoxic venom 230
New Jersey light-trap 11
Nile monitor xvi, 180
northern curly-tailed lizard 171
nuisance alligator program 139

oak toad 44
Ocypode quadrata xv
Oklahoma State University 193
Oklahoma State University xxi
Ordinance 75-29 141
Orianne Society xix, 27, 251
ornate diamond-backed terrapin 118
Overseas Railroad 144
oviparous 184, 201
ovoviviparous 190, 205

Padre Island National Seashore 116
Parker, Daniel xix, 267
parotoid gland 73, 246
parthenogenetic 246
passive introduction 36
peninsula,
 cooter 124
 ribbonsnake 224
Periwinkle Park and Campground 43, 165
pest-plant control operations 17
pet trade 144
Phillips, E.J; Ed xvii, 87, 88, 98
Phillips, Steve xviii, 31, 46, 221
pig frog 37, 64
Pine Island 27, 182
Pine Island Sound 2, 3, 14, 22

Pine Island Sound Eastern Indigo Snake Project xix, 196
pit viper 212, 230
pitfall traps 31
plant transition 17
Point Ybel 6, 22, 115, 116, 126
Porter, Heather xix
Portuguese-man-of-war 79
prescribed fire 26
Presidential Proclamation 2758 8
Procambarus alleni 253
Procambarus fallax 253
Procyon lotor 81
Puerto Sur Nivel 4
Punta Rassa 5
Python Challenge-2013 236

rabbit,
 cottontail 189
 Florida marsh 188
raccoons 129
racerunner, eastern six-lined 157
rain frogs 54
Rancho Nuevo 116
ratsnake,
 red 217
 yellow 214
rattlesnake,
 dusky pigmy 232
 eastern diamond-backed 14
red,
 cornsnake 217
 mangroves 10, 17
red-eared slider 131
Redfish Pass 4, 13
Rhacodactylus leachianus 255
Rhizophora mangle 10, 17, 20, 142

ribbonsnake, peninsula 225
River, Estuary and Coastal Observing Network 103
roadkill data 32
roadkill surveys 28, 31, 32
Rogers, Maggie May 119
Rossman, Douglas xvii
Rutland, Clarence 191

salamanders 39
salinity 24, 25, 39, 67, 120
saltwater intrusion 25, 43
San Carlos Bay 2, 5, 22, 66
Sanibel,
    Causeway 22
    City Council 141
    National Wildlife Refuge xxi, 7, 30
    Nile Monitor and Green Iguana Management Plan 170
    Police Department 140
    10, 43, 91, 94
    Slough 9, 10, 11, 16, 18, 32, 43, 136, 208, 225
    tomatoes 6
    Wildlife Committee 28
Sanibel Island Light Station 6, 10, 179
Sanibel-Captiva,
    Audubon Society 82
    Conservation Foundation 7, 31, 98, 140, 170
    Islander 87
    Mosquito Control District 11
    Road 27
Sanybel Island 4, 6
Sawicki, Michael 161
scarlet kingsnake 197, 199
Schinus terebinthifolius 17, 108, 203
scientific names 36
Sciurus carolinensis 189
sea turtle,
    conservation 117
    green 86
    hawksbill 102
    Kemp's ridley 116
    leatherback 97
    loggerhead 82
Sea Turtle Stranding and Salvage Network 103
Sebastian Inlet State Recreation Area 99
Second Seminole War 6
Serage-Century, Dee 146
sex determination 97
Shell Mound Trail 202
Sigmodon hispidus 189
skink,
    little brown 178
    southeastern five-lined 175
slider,
    red-eared 132
    yellow-bellied 132
snake,
    black racer 183, 184
    coachwhip 185
    indigo snake 26, 30, 187, 195
    mangrove saltmarsh 20
    southern ring-necked 193
snapping turtle 89
snook, common 5
Sobczak, Charles 161
Society for the Study of Amphibians and Reptiles 35, 252
Solenopsis sp. 202
southern,
    black racer 183, 184
    chorus frog 60
    cricket frog 41
    leopard frog 68
    ring-necked snake 193

toad 47, 48
Southwest Florida Regional Alligator Association 139
species 36
species recruitment 23
spring tides 22
squirrel treefrog 55
Stanbury, Fred 139
Stewart Pond 17
storm-surge flooding 5
striped mud turtle 109
subspecies 36
surface water aquifer xvi
Sweet, Kyle 112, 173
Sylvilagus floridanus 189
Sylvilagus palustris paludicola 189
sympatrically 247
Syringodium filiforme 85
Systema Naturae 35
systematic zoology 36

Tabbachi, Marshall 133
Tarpon Bay Weir 43
terrapin, ornate diamond-backed 20, 120
Texas coralsnake 199
Thallasia testudinum 85
Thompson, Richard xvii
tidal amplitude 21
tidal curves 21
Tidal Flats Zone 20
tidal rhythms 21
tide,
    diurnal 21
    mixed 22
    semidiurnal 21
toad,
    cane 72

eastern narrow-mouthed 52
oak 45
southern 48
Wyoming 70
tortoise, gopher 158
tortoiseshell 100
toxic runoff 63
treefrog,
    barking 58, 252
    Cuban 59, 185
    green 54
    pine woods 56, 252
    squirrel 56
Tritaik, Paul xviii, 145
Truman, Harry S. 7
trunkback turtle 95
Turkey Point Nuclear Power Plant 142
turtle excluder devices 81
turtle grass 85
turtle,
    Florida box 128
    Florida chicken 93
    Florida mud 112
    Florida softshell 76
    snapping 90
    striped mud 110

U.S.,
    Army Corps of Engineers 23
    Bureau of Land Management 67
    Customary Units of Measurement 36
    Fish and Wildlife Service 7, 70, 170, 253
    Lighthouse Establishment 6
Upper (or North) Captiva 3, 27
Upper Respiratory Tract Disease 108

vocal sac 65

voucher specimen(s) 37

water hyacinths 23, 24
watersnake,
    Florida 207
    Florida brown 213
    Florida green 210
    mangrove saltmarsh 205
Weber, Brian xviii
West African rainbow lizard 149
West Government Pond 67, 138, 145
West Impoundment 67, 208
Westall, Mark "Bird" xviii, 139
Weymouth, George xvii, 116, 138, 139, 145, 165, 198, 254
Wheeler, Sheena 74
white mangroves 19
Wildlife Committee 233, 252

Wildlife Drive 79, 120
Williams, James 191
Wood, W. D. 10, 66
Woodring Point 22
Woodring, Ralph xviii, 87, 103, 219

yellow ratsnake 216
yellow-bellied slider 130
Young, Victor xix
Youth Conservation Corps 202

Zocki, Roger 145
Zostera marina 85